And I thought I knew QTP!

QTP Concepts Unplugged

By Tarun Lalwani

Publisher: KnowledgeInbox
Technical Editor & Reviewer: Anshoo Arora
Editor: Chhanda Burmaan
Illustrations by Jophy Joy
ISBN: **978-0-9836759-0-7**

© 2011 KnowledgeInbox. All rights reserved.

Printing History:
September 2011: First Edition

Copyright © 2011 Hewlett-Packard Caribe B.V. Reproduced with Permission

No part of this publication may be reproduced, stored in a retrieval system or transmitted in any form or by any means, electronic, mechanical, photocopying, recording, scanning or otherwise, except as permitted under Sections 107 or 108 of the 1976 United States Copyright Act, without either the prior written permission of the Publisher, or authorization through payment of the appropriate per-copy fee to the KnowledgeInbox. Contact online at **KnowledgeInbox.com/contact-us**.

Source codes discussed in this book can be copied, modified or distributed without permission from author/publisher by including the below reference comment header.

```
'Source code taken from "And I thought I knew QTP! - QTP Concepts
Unplugged" By Tarun Lalwani'
'The link for downloads is KnowledgeInbox.com/demos/
QTPConceptsUnplugged_SourceCodes.zip'
'Website: KnowledgeInbox.com/books/qtp-concepts-unplugged'
```

This document also contains registered trademarks, trademarks and service marks that are owned by their respective companies or organizations. Publisher and the author disclaim any responsibility for specifying which marks are owned by which companies or organizations.

LIMIT OF LIABILITY/DISCLAIMER OF WARRANTY: THE PUBLISHER AND THE AUTHOR MAKE NO REPRESENTATIONS OR WARRANTIES WITH RESPECT TO THE ACCURACY OR COMPLETENESS OF THE CONTENTS OF THIS WORK AND SPECIFICALLY DISCLAIM ALL WARRANTIES, INCLUDING WITHOUT LIMITATION WARRANTIES OF FITNESS FOR A PARTICULAR PURPOSE. NO WARRANTY MAY BE CREATED OR EXTENDED BY SALES OR PROMOTIONAL MATERIALS. THE ADVICE AND STRATEGIES CONTAINED HEREIN MAY NOT BE SUITABLE FOR EVERY SITUATION. THIS WORK IS SOLD WITH THE UNDERSTANDING THAT THE PUBLISHER IS NOT ENGAGED IN RENDERING LEGAL, ACCOUNTING, OR OTHER PROFESSIONAL SERVICES. IF PROFESSIONAL ASSISTANCE IS REQUIRED, THE SERVICES OF A COMPETENT PROFESSIONAL PERSON SHOULD BE SOUGHT. NEITHER THE PUBLISHER NOR THE AUTHOR SHALL BE LIABLE FOR DAMAGES ARISING HEREFROM. THE FACT THAT AN ORGANIZATION OR WEBSITE IS REFERRED TO IN THIS WORK AS A CITATION AND/OR A POTENTIAL SOURCE OF FURTHER INFORMATION DOES NOT MEAN THAT THE AUTHOR OR THE PUBLISHER ENDORSES THE INFORMATION THE ORGANIZATION OR WEBSITE MAY PROVIDE OR RECOMMENDATIONS IT MAY MAKE. FURTHER, READERS SHOULD BE AWARE THAT INTERNET WEBSITES LISTED IN THIS WORK MAY HAVE CHANGED OR DISAPPEARED BETWEEN WHEN THIS WORK WAS WRITTEN AND WHEN IT IS READ.

Dedication

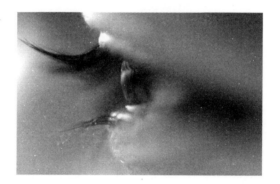

Eyes are a gift of God to us, to be able to see how beautiful this world is. But there are few souls out there who are not blessed with this Gift. I would like to dedicate this book to those who are visually impaired or blind, and have to work harder for every day-to-day challenge. I promise to donate 2% of the profits generated from this book to institutes helping visually impaired people.

Foreword

ELFRIEDE DUSTIN

In recent times, many IT companies have realized that software testing needs to be an integral part of their R&D efforts. Companies on the leading edge of software development, such as Google or Facebook, recognize the need for automated testing approaches. For example, Facebook[1] *"is safely updated with hundreds of changes including bug fixes, new features, and product improvements. Given hundreds of engineers, thousands of changes every week and hundreds of millions of users worldwide,"* Facebook relies on their automated testing program that includes unit and water (GUI) testing as part of their release efforts. Another example, Google[2] *"uses a product team that produces internal and open source productivity tools that are consumed by all walks of engineers across the company. They build and maintain code analyzers, IDEs, test case management systems, automated testing tools, build systems, source control systems, code review schedulers, bug databases….The idea is to make the tools that make engineers more productive. Tools are a very large part of the strategic goal of prevention over detection."*

The message is clear: quality software can't be released without an effective automated testing program. Numerous automated testing tools are available to support these efforts and one of the most popular 3rd party automated testing tools is HP's QuickTest Professional[3] : *"HP continues to be the dominant player in the market, with a presence in virtually every large enterprise. …. Its position requires all other players to position around HP's tools, and is strong enough that several competitors also have integration to HP products. Virtually all SIs, outsourcing providers and testing consultancies support the HP product line, making it easy for organizations to find experienced testers. SAP now resells HP testing tools as part of its overall quality solution. …HP has grown its breadth in the quality space through acquisition, adding strong offerings for security analysis. New product releases in 2010, and planned for 2011, show that HP has retrenched in technical innovation to extend the breadth of its quality solutions (adding test data* management and manual testing) and participate in a wider portion of the ALM market. ... The company has a broad set of tools for software quality, including: Functional Testing (including QTP)," among many other tools.

A dominant tool such as QTP needs to be accompanied by a "How to.." book. Tarun Lalwani has taken an encouraging lead in providing valuable QTP information to help educate new and experienced QTP users, with his first book '*QuickTest Professional Unplugged*' and now followed it up with '*And I thought I knew QTP!*' With this book, Tarun has come up with a 'must-read' guide that can help a user implement successful QTP testing programs and efforts. Even though the book is about a dry topic such as QTP automation, it is well written and entertaining and in the format of an 'interview'. It gives a thorough dissection of many possible QTP problems via various QTP questions and answers. This book articulates solutions to most QTP issues a user will run into and provides timely and necessary material for any QTP user – beginner or advanced.

Elfriede Dustin
Automated Software Testing Evangelist
www.idtus.com

Elfriede Dustin, an IT veteran with over 20 years of experience behind her, has authored and co-authored various books on Software Testing like 'Quality Web Systems', 'The Art of Software Security Testing', 'Automated Software Testing', 'Effective Software Testing' and 'Implementing Automated Software Testing'. She currently works for ITD.

 1 https://www.facebook.com/Engineering#!/video/video.php?v=10100259101684977&oid=9445547199&comments

3 http://www.gartner.com/technology/media-products/newsletters/itko/issue1/gartner.html

 2 http://googletesting.blogspot.com/2011/01/how-google-tests-software.html

Foreword

AJ ALHAIT

"And I thought this was going to be just another technical book!"

Tarun's new book "And I thought I knew QTP!" is a unique revolution in technical book authoring! This was unlike any technical book I've ever read. The approach is so interesting, it makes you want to read on like you are reading a sci-fi novel. It's quite interesting and a fun read, but do not expect to use it as a technical reference manual. This is one book that you can actually learn something from by reading it start-to-finish.

Tarun has, in a way, revolutionized technical book authoring by writing this book... I expect others to 'get inspired' and use a similar structure and style in the future. Kudos to Tarun for writing this 'out of the park' hit.

AJ Alhait
Founder

 SQAForums.com

QAtraining.net

Amjad Alihat (AJ), founder and owner of SQAForums.com and QAtraining.net, has over 18 years of experience in software testing, and has been working as a consultant for over 10 years. Started in 1999, SQAForums.com is a hub for various QA professionals and has around 190,000 members. As a knowledge-sharing platform it has helped many professionals to increase their expertise. Tarun also started his knowledge sharing through SQAForums, later expanding the same by launching his blog KnowledgeInbox.com and now through his books.

Preface

While conducting job interviews recently for my company, I found that many candidates failed to answer simple questions on QTP. I could easily gather that some of them had practical experience but failed to explain the concepts behind it. Whereas some candidates were obviously exaggerating their level of knowledge or expertise in the given field.

Until a year back, my first instinct on any public forum was to refrain from answering any interview questions, so as not to promote or encourage standardized answers or discourage independent thought. Even if I did answer, I would first check to see what efforts the person had taken on his/her own to look for the answer.

But recently, a mock interview with a teammate started a chain of thought. During the interview, she got nervous and was unable to answer some basic questions, even though her practical knowledge is quite strong. It was then that I realized that people may have practical experience or expertise and still lack conceptual knowledge of QTP.

After this incident, I started looking at various forums only to find QTP questions being asked and their responses. The worst part of this exercise was to see so many incorrect answers being offered by novices or inexperienced users. For example, one of the questions, 'Is it possible to use JavaScript in QTP?' was answered as, 'Yes, we can use JavaScript in QTP. But before that, we have to install Java Add-in'. This is only one of the replies among many that may spread misinformation amongst the beginners. There were other users who replied in support of the above answer. Even though I was surprised to find such replies and support, I was, at the same time, a little concerned about incorrect knowledge being shared by users like this.

Today, experts worldwide refrain from answering such queries with the intent of discouraging standardized answers. But my recent experience made me realize that it is even more important for experts to answer these questions correctly in order to eliminate the ever increasing gap in conceptual knowledge of QTP.

With this goal in mind, I decided to write a book on QTP with a fluid storyline and a dialogue-based approach instead of a plain FAQ style format (as it limits the scope of dialogue) that most technical books seem to follow. The result is a book which hopefully makes learning basic-to-really-complex QTP concepts interesting and entertaining. It is my hope that I have achieved (atleast to some extent) what I had set out to do when I started writing this book; that is *eliminate misinformation and doubts in the minds of QTP practitioners*.

Target Audience

Manual Testers, Analysts, and/or Managers who want to switch to Automation or QTP users with any level of expertise should benefit from this book. The book touches on various concepts of QTP and attempts to provide information that is missing in theoretical and practical domains of QTP. Attempt has been made to explain concepts in a simple way without sounding too simplistic.

All characters/names appearing in this work are purely fictitious.
Any resemblance to real person(s)/company/names/material/product is purely coincidental.

December 2010

ॐ

*M*y flight landed at the Pune airport around 11:00 a.m. One thing I always loved about the QueenFisher flights was their on-time performance. I was tired and eager to pick my luggage and rush back home but a technical glitch at belt #2 delayed that by almost half an hour. I had gotten used to long waits by then. After waiting for 40 odd minutes I finally spotted my bag. I picked up the bag and took a pre-paid cab home.

Strangely it was raining heavily and there was a high probability of getting stuck in a traffic jam. Luckily the situation was not as bad as I had expected and I reached home around 1.30 p.m.

I had planned earlier to go to work for the remainder of the day but after the exhausting flight and subsequent waiting, I changed my plan and decided to take the 2nd consecutive day off. I dozed off for few hours and was in deep sleep when I suddenly woke up with a very strange feeling. I couldn't remember feeling this anxious and nervous in my life. I started biting my nails for the first time in my life wondering what would happen next. I knew I couldn't have done anything differently. I tried convincing myself not to think too much about it and somehow passed the day.

The next day was Friday, a day when the office takes on a more cheerful look as we are allowed to dress in casuals. I reached my desk and opened Outlook to check my emails. When I checked my inbox, it felt like I had been away for ages although I had been on leave for two days only. Two Hundred and Twenty Six unread emails and by the time I finished reading them all I realized two hundred of them were just forwards or mails that weren't of any interest to me. I thought to myself, *'what a waste of time'* but then

And I thought I knew QTP!

suddenly lightning struck me. I realized I was on bench and I would need such things to kill time.

It felt really strange to be on bench after seven years of working on tight delivery schedules. I won't say it felt great as I always preferred to work but I knew I won't mind this break for a few days.

It was time to catch up with my friends and on the lunch table they had only one question, 'Where were you for the past two days?' I somehow managed to change the topic and avoid answering their queries. They knew I was up to something but they had no idea what.

I spent the rest of the day reading some articles that I had saved on my desktop. Some of the articles were so engrossing that hours passed by before I realized that it was almost time to go home.

On reaching home I decided to watch my favourite movie on DVD — 'The Matrix'. I just loved the concept of the movie. The most interesting question the movie raised was regarding our interpretation of reality.

I was already halfway through the movie but at my back of my mind all the events from the past week were being replayed. There was this one thing that I knew might change the whole game and I was afraid that I might have got it all wrong. But I knew there was nothing I could do now as it was a thing of the past. If it was just one person's decision then I knew it would have been in my favour but there were many people involved in this.

My roommate had left for his home town and would be away for two weeks. I knew it would be tough to get the weekend through on my own, so I decided to call up few of my friends and go for a movie. All of us met at FSquare and enquired about the current shows. To our disappointment all movie shows were running full. I wondered if all the people in the city had decided to watch a movie on that day.

December 2010

Suddenly we noticed a guy giving away free tickets for the 'Feel N Freaky' movie. I knew the movie was pathetic but then we had no other options as well. We asked the guy about the offer and he told us that if we had a KodaFone cell connection we can get two tickets free. Three of us had that cellular service provider and I thought it was our lucky day. Alas, it was not to be. He just had the last two free tickets to offer. Since we had no use for just two tickets, we decided to try our luck at another movie theatre nearby.

My friend asked me, 'What about these tickets?' to which I replied, 'Nothing, let's just throw them.'

Lalwani

And I thought I knew QTP!

While crossing the road on the way to the car parking, I saw two girls approaching from the other side. On seeing them, I got an idea. I walked up to them and said:

Me: Hey, do you want to watch 'Feel N Freaky?'

(She looked at me with those unsure eyes wondering if it was she I was talking to or someone behind her.)

Girl: Yes, but..

Me: Okay, I have two tickets for the movie if you want. And without even waiting for her reply I handed her the two tickets.

Girl: How much?

Me: It's free. Enjoy the movie.

Girl: No, I would like to pay. And she took out some money.

Me: Don't worry, it's free; I can't watch it.

(I knew that after watching the movie she would be cursing me.:D)

I crossed the road and my friend had already brought the car outside the parking lot. Finally, we went to another theatre and this one had tickets available. We knew we could watch the movie of our choice and we all agreed upon watching 'Inception'. The show was scheduled to start at 6:30 p.m. and it was 5:15 p.m. then. So we decided to head to FarBucks coffee shop and chitchat over a hot cuppa. Relaxed and sipping my vanilla latte, the events of the preceding days came back to my mind again. But suddenly Raju interrupted me and I snapped back to the present.

The movie started at 6:30. To say I was awed would be an understatement. I had never seen any movie in my life that had a concept as great as this. Stepping into someone's dream and stealing thoughts. I really admired the storywriter's imagination and grasp on the topic of dreams. It was just truly incredible and made me realize how little we know about what goes on inside our brain when we sleep. But honestly I was not interested in knowing about it as well; I prefer deep sleep where I don't know what is happening ☺.

After the movie we all went for dinner at InLand China. It was my first visit to this restaurant and I was having Chinese cuisine after a long time. I thoroughly enjoyed the food and felt happy.

I woke up around noon on Sunday and had a lazy brunch. After reading the papers, I got busy with getting the house cleaned.

Monday arrived and I started the day with high hopes. I was waiting for that *one* email.

And I thought I knew QTP!

Since my 'WhyMail' account was blocked in the company, I had set a forwarder to my office email id. The whole day, the only thing I did was to hit the F9 button on my keyboard and scan every new email. But the one I was waiting for never came. This continued for the next 3 days and I had the sinking feeling that I had lost it. I felt very disappointed and dejected too, but I knew this is not the end of the world. I recollected the statement that I had quoted to Andrew.

'Success or failure is a part and parcel of life. You learn from failures and move on to new challenges with even stronger determination.'

Days passed and it was Friday again. I had settled back into to my normal routine. It was around 11:00 a.m. when I got a call from a Delhi number. My eyes lit up as I wondered if this was the call I had been waiting for.

Me: Hello

Caller: Hello Sir, I am calling from ISEEI bank. Would you like to apply for a home loan?

December 2010

Me: *(I nearly lost it! It felt like destiny was playing games with me. I wanted to disconnect the call immediately but I knew she was just doing her job.)*

Yes madam, are you giving any free Home also along with that loan?

Caller: No sir, we can just give the home loan.

Me: But what will I do with the home loan without a home to apply it for?

Caller: Sir, that you will have to find on your own or else we can send our consultants to assist you.

Me: Hmmm, ok how much loan can I get?

Caller: Sir, can you please tell me where you work and what is your monthly income?

Me: I have a small tea shop in CannotPlace, Delhi and it does well but I can't disclose my income.

(By now she knew what I was doing and she just hung up on me.)

After the call ended, I again went back to reviewing all the events of the past few weeks and replaying in my mind the sequence of events that had started on a regular Monday two weeks ago.

Two Weeks Ago

৸৻ঌ

Ahh..What a wonderful Monday! It is rare to be happy at the start of the work-week but this was one such day as the most demanding project I had worked on got over. Eventually, all the late nights and weekend work hours had paid off and it was time to soak in the appreciation from the client. Due to a high-pressure, tight-deadline environment, listening to all sorts of requests, cribbing and scolding had become a norm. Today though, I was looking forward to hearing some sugar-coated words. At work, nothing can make your day better than hearing some appreciative feedback from the client. Suddenly, a pop up appeared on the screen and I knew it's the team meeting call with client. It was 10:00 a.m. and everyone had begun stepping inside the meeting room for the conference call with client.

I jumped into the meeting room and started checking if the whole team was in or not. We had Tiniv, Yogi, Rehm, Jhinga, Bignes and my PM Saili. But Kates was missing. I called her up and asked her to rush for the meeting as only two minutes were left for the meeting to start.

Charles, the project lead from the client side joined in and the meeting started.

I asked everyone to put their phones on the silent mode as Charles hated such interruptions during the con-call.

Charles: Morning everyone!!

Team: Good morning, Charles!!!

And I thought I knew QTP!

Charles: I believe all of you know that this is our final call for this project. It's been six months since we started working on this project but I still feel as if we started just few days back.

(I thought to myself, 'Oh please, Ask us!! It feels like I have been in this project for over a year now.')

First of all, I want to thank all of you for all the hard work you have put in the last six months of this project. I know we have faced many challenges and you have been putting in a lot of extra effort off late to get over them. But finally we have done it.

It would have been impossible to get this done without the dedication that you have shown and I am really proud of each one of you. I really feel honoured to have led this team and I hope I will get a chance to work with you in our future assignments. Saili, would you like to add anything?

(I just couldn't believe that it was Charles talking. It seemed as if he was a totally different person today or maybe he had sent someone who could mimic him. All our daily calls with him used to be so different, with the team being pushed for issues which had nothing to do with our team. Our work used to be stuck because of issues from other team and that made sure we lost our weekends.)

Saili: Yes Charles. First we would like to thank you for all the support you have been providing. Without the same it would have been difficult to finish the project within the given timeline.

(I looked into Saili's eyes and grimmaced... Support??? And she just smiled back.)

I would like to thank everyone in the team as well. I know that all of you have been stretching yourself for the past few months. We have accomplished a commendable task in this time and I know it would have been impossible without such a great team.

Jingha, you have done a great job in leading the team and managing all the deliverables.

Nurat, thanks for all your technical support and consultancy for this project. Your expertise in QTP made sure we had solution to all our technical issues.

Two Weeks Ago

Rehm, Bignes and Kates. I know this was your first Automation project and you all managed to work in a way as if you had years of experience in QTP. The way you all picked up automation and QTP was just truly incredible.

(I could see smiles all around and I knew what that comment meant for all of them.)

Tiniv, Yogi. I know your module SWAT was the toughest one to understand and automate. But you managed not only to understand the application better but also automate most of its complex scenarios which we earlier had our doubts on.

(The team was in high spirits and most of the folks had decided to go back home for a week's leave, so it was all happiness in the air.)

Charles: Thanks Saili! By the way I forgot to tell you that management is very happy with our project and has decided to gift 8GB IFod to each of you as a token of our appreciation.

(We put the phone on mute for a few seconds as the room suddenly erupted in shouts of joy; the team enjoying the news of this unexpected gift from the client. In my seven years of experience this has never happened before, at times it was even difficult to get an appreciative email.)

And I thought I knew QTP!

Our calls with Charles have been pretty stressful in the past but this kind gesture more than made up for it. Un-muting the line, we all thanked Charles.

Team: Thanks Charles!!!

Charles: Okay everyone, I have to join another meeting with the senior management, so I guess we need to end it here.

Thanks again everyone and have a great day!

Thirty minutes later, the meeting was over and we all stepped out of the meeting room. I was heading to my desk and took out the eyePhone only to find '3 missed calls' from an unknown number. Before I could check it out, my colleague Raju tapped on my shoulder from behind and said, 'Nurat, my buddy, now no more working on the weekend and no more late nights.'

I smiled and said, 'Yeah, I hope so too. This project was really challenging. Everything that has a beginning has an end.' We both started laughing and I forgot about the call and put the eyePhone back in my pocket.

'Let's go for coffee. What say?' Raju offered. I thought for a moment and accepted saying, 'Why not, we are on bench for today at least, don't know about tomorrow.' I called up my friend Kulu as well and asked him to join us.

We walked out of our building and took a turn towards the coffee shop. Apparently, the thought of the missed call kept bugging me. In the meantime we reached the coffee shop and Raju ordered three cappuccinos. Kulu also joined us shortly thereafter. We settled into a table in the corner and soon our coffees arrived. Raju took a sip and broke the silence, 'So Nurat, what next? Will you stay here or have you started looking outside for a good opportunity?'

Two Weeks Ago

I also took a sip and said, 'Not really sure, but now I want to do more challenging work on QTP.'

To that Kulu asked, 'How many more challenges do you want? Wasn't this project enough for you?'

'He always loves to work on QTP; it seems he wants to master it.' Raju jumped to my defense.

'Naa… Nothing like that. You know it's not about knowing QTP and its capabilities. But it is more about how to use them effectively in a given situation. So the tougher the projects I take, the more chances I have on learning the implementation part of it,' I concluded.

And I thought I knew QTP!

Soon we were done with our coffee and returned to our desks. The moment I put the eyePhone down on the desk, it began to blink. I realized I still hadn't put my phone back to the normal mode after the meeting. I picked up the call and it was my Mom.

Me: Hi Maa (Mom)!

Mom: Hi Nanu, how are you?

(My mom used to call me Nanu. I hate it when anyone, except her, uses that name. I thought for a second, that she generally doesn't call me at this time…)

Me: Maa, I hope everything is alright?

Mom: Yes everything is fine, don't worry. It's just that I have received your new debit card and I wanted to let you know the same.

(I had requested for a new card as my old one stopped working at most of the ATMs.)

Me: Oh great, Maa you keep the card I will collect it when I come down to Delhi next time.

(Just then, I heard the beep indicating I had another call waiting.)

Me: Maa, I am getting another call, I will call you from home tonight.

With that, I picked up the second call.

The Consultant Call

Me: Hello

Caller: Am I talking to Mr Nurat?

Me. Yes, speaking

And I thought I knew QTP!

Caller: Hi Nurat, I am Megha calling from No Pay Consultancy.

Me: Ok?

Megha: I found your profile on BeastJobs.com and one of our esteemed client has some requirements in Test Automation.

Me: Ok, what type of role is it?

Megha: Our client is developing a framework on QTP and is looking for people with expertise in framework design and development.

Me: Sounds good. May I please know who the client is?

Megha: It is MecroHard. They have offices in US and India, but this opportunity is for their offshore centre.

Me: This role seems interesting to me.

(I just couldn't believe that I had a possible opportunity in a company everyone dreams to be a part of.)

Megha: So can I arrange a telephonic interview for you?

Me: Yes, fine. But when would it be?

Megha: Their requirement is a bit urgent, so the telephonic interview will happen today itself. Would 8 p.m. today be fine with you?

Me: Fine.

(I thought I would leave on time today, anyway since I was on bench from today it shouldn't be an issue).

But what kind of interview would it be?

Megha: It will be a technical interview and for now they will only check the basic QTP concepts. So just to confirm, I am scheduling your telephonic interview for tonight at 8 o'clock.

Me: Yep, it's fine.

It was 11:30 a.m., so I thought of catching up with a few friends for lunch and leave office early by around 5 o'clock. I wasn't much worried about the interview as the consultant had said it would be QTP basic interview and I have been working on QTP since a long time. However still, a part of my conscience gave me a worried pinch at that particular time interval as if someone had set a clock inside my head.

I called Raju for lunch along with Kulu and Uma. We met in the cafeteria and seized a table somewhere in the middle this time. After quite a long time we all were having lunch together but my project was topic of discussion here also. Except for Raju and me, all were in manual testing. Hence it was understandable that they were quite excited about knowing QTP.

After lunch I reached my desk and unlocked my computer and found an unread mail from Mr. Prasad saying, 'Come and meet me.' Thinking it was about a new assignment I went to his cubicle where he was staring at the screen of his laptop with complete concentration. I stood there for couple of moments before drawing his attention towards me by saying, 'Mr. Prasad, you called me?'

'Oh yeah Nurat, actually I want you to prepare the learning note for this project.' said Mr. Prasad. It was more of an instruction than a request. 'Oh, ok,' I mumbled and went back to my desk. I was thinking to myself why Saili didn't ask me for this and why Prasad was involved now. But I knew whatever it was I had to prepare that document. While preparing the learning document for the project I came across so many new things about QTP and that we had learned in this project. Initially, I thought of completing the document preparation by 4 o'clock or at the latest by 4:30 p.m. but when I looked at my wrist-watch after finishing the document it was 5:30 p.m. I was amazed at recollecting all the new things that I had learned during this project. My confidence for the upcoming interview went one notch up.

And I thought I knew QTP!

As I wanted to be home on time, I left office quickly and was hoping to avoid the buses leaving at 6 p.m. But being a cautious driver I knew I was hoping against hope. It was impossible now to avoid the peak-hour traffic jam, and today was worse than usual. I felt that I could reach faster if I had paddled on a bike. The traffic was inching forward very slowly much to my increasing panic. To make matters worse, few vehicles decided to take on the wrong side of the road and jammed up that road as well.

I just could not believe that on the day, when it was important for me to reach home on time, I was caught in this mess of a traffic jam. It was 6:30 p.m. and I had only covered a distance of 1.5 kms from office. I felt irritated but I knew it was not gonna help much. Finally around 6:50 p.m. a few traffic policemen showed up and the traffic started moving slowly. What usually used to be a 25-minute trip took an hour and a half that day.

The Consultant Call

I reached home and decided to take a hot water shower. The interview was still 40 minutes away and I wanted to relax my mind before that. Finally it was 8:00 p.m. and I sat waiting for the call. There was pin drop silence in the room and I could hear the clock ticking away. It was 8:15 p.m. and it felt as if with every passing second time was getting slower and slower.

Finally, it was 8:40 p.m. and by now I was very hungry. I thought to myself that may be the call was not going to happen now, so I decided to go down and have something for dinner. The moment I stepped outside the door my eyePhone started ringing. It was as if I had stepped onto some kind of sensor. I went back inside and picked up the call.

The Telephonic Interview

Me: Hello!

Caller: Hi Nurat, this is Ekta from MecroHard. Sorry about the delay. I will now put you through to the person who will take your interview.

Me: Okay.

Interviewer #1: Nurat, hi, I am Samir. Please accept my apologies for not being able to call you on time. Something urgent came up. Since we're already behind schedule I will not waste more of your time and start with the interview right away, if that is okay with you.

Me: Sure.

Samir: Can you brief me about your experience?

Me: Sure. I am with Sysfokat for over 7 years now and out of this, I have worked in manual testing for one year and the remaining six years in Test Automation using QTP and BPT.

Samir: Ok. Tell me what is Automation?

Me: (What? What kind of a question is that? Never heard anyone asking that. It's hard to explain in words...mmm....)

In simplest words, Automation is the process of removing manual intervention from a given task.

Samir: What tools are you aware of that are used for Automation?

Me: I work primarily with HP's QuickTest Professional. But, there are other tools for Automation as well which I haven't had a chance to work with. To name few – Selenium, Rational Functional Tester, Ranorex, SAHI and AutoIt.

Samir: You said you have done manual testing for one year. Tell me what different phases you had in it and what were the main activities?

Me: In manual testing, our work used to be divided in core phases of Planning and Execution. In the planning phase, we used to study the requirements, capture atomic test conditions, create test scenarios, create test cases and then upload them in QC.

In the execution phase we executed the test cases, logged defects, created daily and weekly summary reports and also the final closure reports.

Samir: So you also must have logged few defects in QC?

Me: Yes, I did.

Samir: What fields were mandatory?

Me: I may not remember all of them but I will list as many as I can: Detected By, Assigned to Team, Assigned To, Reproducible, Description, Priority, Severity, Application Name, Release Name and Status.

Samir: You mentioned Severity and Priority. What is the difference between these two and give me an example where these 2 were different?

Me: Severity for us was based on the defect's functional impact on the application while priority was based on the business impact of the issue.

For example — If there is a 'Terms and Conditions' popup with a legal statement missing, then this does not have a functional impact on the application. So, we would assign it a Low Severity defect. But since this involves legal text and should be fixed ASAP, the priority chosen would be 'High'.

Samir: What type of testing did you do?

Me: Type as in?

Samir: I mean integration, system etc.

Me: I was into System testing.

Samir: What is the difference between Integration and System Testing?

Me: It is in terms of depth of testing. Integration testing has small set of scripts which just test if the application as a whole is working fine and whether different interfaces can interact with each other correctly. System testing on the other hand is more concerned about testing the application's functional requirements and encompasses more scenarios than integration testing.

Samir: What were the next steps after after your testing?

Me: Performance testing used to start in middle of system testing. After our team's approval, User Acceptance Testing and Usability testing were also done by the client.

Samir: How did you trace mapping of your test requirements?

Me: We used to have a RTM (Requirements Traceability Matrix). It mapped all business and technical requirements. We used to map our test case IDs to these technical requirements to make sure everything is covered.

The Telephonic Interview

Samir: Did you face any situation where you had bugs that couldn't be reproduced? If yes, how did you handle them?

Me: Before logging any defects, we always confirmed the issue on the application. In case it worked correctly, we just took a note of any timestamp information. If the issue appeared again, we logged a defect and associated it with the old and new timestamp information. This way one issue was resolved in our application by the help of log files and timestamps.

Samir: What is the maximum number of Actions we can create in QTP?

(Wow! What a fast switch...)

Me: I don't know. But I guess it may be the same as the limit of sheets in an Excel file. I don't exactly remember the count but I guess it was 255. Personally though, I feel that if you reach a point where you have to create more Actions than the limit allowed by QTP, then you are probably looking at a very poorly designed script.

Note: For QTP 8.2 and below the limitation is 255 Actions. For QTP 9.0 or higher the limit is 120 Actions. For more details refer to the KM178776 document on HP support.

Samir: How does QTP identify an object while recording?

Me: QTP identifies objects based on three properties, which are Mandatory, Assistive and Ordinal identifiers. When QTP learns a new object, it captures all the Mandatory properties of the object and tries re-identifying the object. If it is able to find a unique match it stores these properties. If multiple matches are found, then QTP starts adding assistive properties one by one until a unique match is found. If all the mandatory as well assistive properties have been exhausted and QTP still does not find a unique match, it determines the value of the ordinal identifier for the object and adds it as the identification property.

Samir: What types of ordinal identifier does QTP support?

Me: QTP supports three types of ordinal identifier named Index, Location and CreationTime. Index is based on how the object appears internally in the application; location is based on the object's screen position. CreationTime is only available for Browser object. Regardless of the type of ordinal identifier used by an object, it is always a unique numeric value. In other words, at any given time, two objects with the same description cannot have the same ordinal identifier.

Samir: What is the default ordinal identifier used by QTP?

Me: It depends on the Add-in. For all windows objects the default ordinal identifier is Location, while for Browser it is CreationTime and for all other Web objects it is index.

Samir: I add an object to the object repository and it uses 3 mandatory properties and 2 assistive properties. Now when I run my script, which properties will QTP use for identification? Will it use the 3 mandatory properties first and if there is no unique match then it would use the Assistive properties or how exactly would it work?

Me: QTP will use all 5 properties for the identification of the object and there is nothing like mandatory or assistive property in run-mode. The only concept that matters at run-time is the object's defining properties and/or its ordinal identifier.

Samir: What are the different data types supported by QTP?

Me: QTP's scripting language is VBscript, so it supports data types from VBScript. VBScript only supports Variant data types. A variant type can store different types of data like int, single, double, string, object, long, date etc.

Samir: How do you know what type of data is stored in a variable?

Me: We can use the TypeNamemethod which returns the type. We can also use VarType, which returns a constant mapping to the various variable types.

Samir: How you do check if the value is an array?

Me: We can use the IsArray method of VBScript. An alternate way is to check if VarType of the variable is greater than vbArray constant.

Note: Value of vbArray is 8192.

```
arrArray = Array("John", "Mary")
If VarType(arrArray) > vbArray Then
   'arrArray is an array
End If
```

Samir: Can I run my QTP scripts without installing QTP?

Me: No, QTP has to be installed on the machine to run the script.

Samir: I have QTP installed on my machine and I create a script in QTP. Now, I move all of that code to a VBS file and run that from desktop. Will it work?

Me: No, it won't. VBScript is just a language and a Host is required for executing it. When we run the code in VBS, Windows picks the default host which is WScript, which does not understand any of the QTP's Test Objects. Therefore, at the first instance of QTP's Test Object in the executed code, the script will fail. So, it won't work.

Samir: What is the scope of a variable in an Action?

Me: The scope of any variable declared in an Action is limited to the Action's iteration. Once the Action iteration ends all the variables are destroyed.

Samir: What is the difference between object repository and object repository manager?

Me: Object Repository is used to view all local objects present in the current Action and also the shared objects available. Object repository manager on the other hand, is an inbuilt tool used for creating and editing shared object repositories.

Samir: I have launched Object repository manager and opened a shared object repository. The button for adding objects is disabled. Why?

Me: Whenever we open an object repository, it is opened in read-only mode by default. To edit objects in the object repository, we must enable editing first.

Samir: When we add an object to the OR the hierarchy is different from what the object spy shows, why so?

Me: QTP only keeps the part of the hierarchy necessary for it to re-identify the object. Objects like WebTable etc are always ignored when identifying child objects because it helps remove some dependency on script implementation. If the object identification properties remain the same, then QTP will re-identify the object even if some of its parent objects might have changed at run-time.

Samir: I am trying to add an Object to the OR but it doesn't get added, what could be the reason?

Me: Some objects are dynamic in nature and they are destroyed as soon as focus is taken away from them. Such objects are difficult to capture in OR. In such cases, one can spy, note down the properties and then manually create the object in the OR. This issue can still happen for other reasons as well which are mostly QTP related.

Samir: What is the difference between GetROProperty and GetTOProperty?

Me: GetROProperty is a QTP's method to retrieve the value of a property from the actual runtime object whereas GetTOProperty is to retrieve the identification properties we used. In other words, GetTOProperty fetches the value from the OR while GetROProperty fetches it from the actual object. For GetROProperty to work, the object needs to exist in the AUT while GetTOProperty can work even if the object is not present on the screen.

Samir: What is the use of SetTOProperty method?

Me: It is used to update any of the identification properties. This is helpful when we would know the value of that property at run-time. Then, we can change identification properties at run-time itself.

Samir: What is the use of SetROProperty?

Me: There is no such method called SetROProperty. To change RO properties we need to use methods which change them internally like Set, Select, etc.

Samir: I have a QTP script Can I call this complete script from other script?

Me: No, QTP only allows calling Re-usable Actions and not the whole script.

Samir: What are the licensing models of QTP?

Me: There are two models available. One is seat/node locked license and the other is a floating license. Seat/node locked licenses are for a single pc and they need to change in case something changes on the machine like network card, hard drive etc. Floating licenses are installed on the license server and then client machines can specify the license server IP.

Samir: How are CheckPoints stored with the test?

Me: QTP stores CheckPoints within the Object Repository file itself. They are stored in a HP's proprietary binary format.

Samir: What are the types of recording that QTP supports?

Me: Context Sensitive recording, Low-Level recording and Analog recording.

Samir: How do these modes differ?

Me: In Context Sensitive recording mode, QTP will try and identify object at specific granularity E.g. WinEdit, WinButton etc. In Low-level recording, QTP uses generic objects like Window and WinObject. Analog recording will record all mouse movements as well, for replay; this is useful when we are recording on an application like MSPaint.

Samir: What is the default recording mode?

Me: It is the Context-Sensitive mode.

Samir: What is the shortcut for these recording modes?

Me: I have rarely used recording in QTP and I just remember that for normal recording it is F3 and the other two would be based on CTRL, SHIFT or ALT combination keys with F3.

 Note: Shortcut for Analog Recording is Shift + Alt + F3 and for Context Sensitive recording is Ctrl + Shift + F3.

Samir: What are the different types of Actions in QTP?

Me: We have Re-usable Actions, Non-reusable Actions and External Actions which are basically Re-usable Actions called in from other scripts.

Samir: What is the difference between Re-usable and Non-reusable Actions?

Me: Call to Re-usable Actions can be inserted again within the same Test or other Test. But, Non-reusable Actions can only be used once in the Test and cannot be called by External Tests.

Samir: Is it possible to add an Action to a Test programmatically?

Me: We can add an Action to a Test using QTP AOM. But there are two limitations: the minimum version required for this is QTP 10 and this can only be done at design-time. What I mean is that, it is not possible to add Actions through any scripting methodology during script execution.

Samir: Can a single QTP Action have multiple Local DataTable sheets?

Me: No. An Action can only have one local sheet. But we can still add extra sheets at run-time and access them using the DataTable object.

Samir: Okay, can a single QTP Action have multiple Global DataTable sheets?

Me: No. It's one and the same thing. My last answer holds true for this one also.

Samir: Can a QTP test have multiple Local DataTable sheets?

Me: *(He loves DataTable I guess....)*

Yes, based on the number of Actions it can have multiple Local DataTables. But in actual, a Local DataTable is not held at the Test level, instead, it is held at the Action level.

Samir: Can a QTP test have multiple Global DataTable sheets?

Me: No, a Global DataTable can only have a single instance within any QTP Test.

Samir: What is the difference between a design time and run-time DataTable?

Me: Design time DataTable means the DataTable of the Test when test is in idle mode. In other words, a design-time DataTable is the one that is created when the test is not running. Run-time DataTable is the copy of the design time DataTable which includes any changes done at run-time through the script. Run-time DataTable is only available in Test Results summary.

Samir: Can the run-time data be written to Design time DataTable?

Me: No, we can't overwrite design-time DataTable through automation. As I stated earlier, all changes made to the DataTable at execution-time or run-time will be made only to the RunTime DataTable.

Samir: What are the three formats in which QTP report can be generated?

Me: I am aware of only two: XML and HTML.

Samir: What about PDF?

Me. The report can't be generated as PDF but you later export it to PDF. So, technically QTP does not create the report in PDF format out-of-the-box.

Samir: Which property of a WebList determines the number of items present in it?

Me: It is the 'items count' property.

Samir: What is QTP Plus?

Me: It used to be an additional installation that was provided with QTP 8.X. After QTP 9, most of its parts were merged into QTP itself by default. Now there is no QTP Plus.

Samir: You must have heard of CreationTime Ordinal Identifier. Can the CreationTime Ordinal Identifier be used for a Windows object?

Me: No. It is only available for the Browser object and doesn't have any meaning for any other type of object.

(I looked up at my watch and noticed that it was 9:30 p.m. I was feeling very hungry by then and was waiting for this to end, but I didn't know when it would.)

Samir: Can we compile VBScript code in QTP?

Me: No. Compiling usually means converting code to executable format but VBScript is an interpreted language which is interpreted at run-time.

Samir: Is VBScript an Object Oriented language?

Me: No, it is not. It is usually called an Object based language for the simple reason that it does allow ability to create classes and their objects. But since it doesn't support polymorphism and inheritance, it is not an Object Oriented language.

Samir: Which Object Oriented features does it provide?

Me: Abstraction and Encapsulation.

Samir: What is Option Explicit?

Me: It is a statement or a directive in VBScript which enforces variable declaration. So, if we use this we must declare each variable, else VBScript would throw an error. Without Option Explicit, VBscript would just declare the variable by itself with Empty value.

Samir: Can I use Option Explicit in the middle of my code?

Me: Well, I have never tried that out but logically it should never be allowed and would raise an error. I say this because, using strict coding standards aren't generally for part of the code – they're for the entire library.

 Note: Option Explicit always has to be used as the first statement in a Library or VBScript code. Using it in between the code will generate an error.

Samir: Thanks Nurat, I am done here. Is there anything you want to ask me?

Me: No, I am good for now. Thanks.

Samir: Okay then. Good night.

And he hung up the phone. Though the interview wasn't very tough but it took two long hours and I was damn hungry by then. I decided it's time for a pizza. I called up Daddy John pizza corner and ordered for a home delivery.

Next morning while in office, I got a call again from the consultant.

Consultant Call #2

༺༻

Me: Hello Megha!

Megha: Hi Nurat, how was the interview yesterday?

Me: It went well from my side. Did you get a chance to speak to them?

Megha: Yes, I was on the phone with them just now. They want to do a face to face interview tomorrow.

Me: Tomorrow!! Where would the interview be held?

Megha: It will be held in their Delhi office. The core framework team works from that location.

Me: Can't we do it on a weekend? It would be difficult for me to make travel arrangements from here to Delhi on such a short notice.

Megha: You don't need to worry about that, your to and fro air tickets will be booked today itself once you confirm. You will also be covered for a day's stay in a nearby hotel as well.

(This was happening just too fast for me and I didn't know what to say. But I have no option to say no, do I?)

Me: Well I will still prefer this weekend but if it is not possible then I will travel tomorrow.

Megha: Fine. Just hold on for a minute and I will confirm for you now.

And I thought I knew QTP!

Me: Okay.

(After two minutes)

Megha: Nurat, I will email your e-tickets for tomorrow's travel. And please make sure you carry copies of last 3 months' salary slip, your passport, your final year college marksheet and 10th and 12th standard marksheet as well.

Me: Fine, I will carry those.

I got the tickets in the email and the return ticket was for Thursday morning. I thought I would let mom know that I am coming down to Delhi for the interview. I called her up.

Me: Hi Maa

Mom: Hi Nanu

Me: Maa, I am coming to Delhi tomorrow.

Mom (In a shocked voice): What happened?

Me: Nothing Maa, it is just for an interview.

Mom: When are you going back?

Me: On Thursday morning.

Mom: Why don't you take a leave on Thursday and Friday and stay at home for four days.

(Well I was so excited about the interview that I forgot to think that way. I could have asked for a return ticket on Sunday night or Monday morning may be.)

Me: Oops! It didn't strike me at that time maa. But I don't want to call them up again and ask to reschedule. It won't look good on my part. I will do one thing as soon as I am done for the interview I will come home and morning I will take a taxi from home itself.

Mom: As you wish. And do remind me to give you your debit card as well.

Me: Sure. Okie Maa gotta go now. I will talk to you later.

I had an early morning flight at 7.00 a.m. and I knew I had to leave by 5 o'clock to be there on time.

I boarded the flight and sat on seat no. C11. I always preferred aisle seats as they allow you get out fast. The person sitting next to me looked familiar but I didn't pay much attention to him. Sometime during the flight, the person started taking to me.

Person: Hi, I am Akash.

Me: Hi, I am Nurat.

Akash: Are you are from Delhi or Pune?

Me: Born and brought up in Delhi. What about you?

Akash: I am from Pune. But my wife is from Delhi.

Me: So you must be going to your in-laws place?

Akash: Yeah. What about you? Going back to the family.

Me: (I took a pause to think if I should say the truth or just say family...)

Yeah, I have some function to attend to in the family.

Akash: Your face seems familiar, I have seen you somewhere. Do you work for Sysfokat?

Me: Yes, you also?

Akash: Yes. I work for the FreeS account.

And I thought I knew QTP!

Me: (Suddenly it came to me in a flash. This guy sits on the opposite side of the floor.)

You sit in building A2, 3rd floor?

Akash: Yes, you also on the same floor?

Me: Yes, I work for the SVI group

Akash: My friend, she is a PM in that group. Not sure if you know her or not. Her name is Saili.

Me: (Now this was getting to be too much of a co-incidence. I was so glad that I didn't open my mouth about the interview else it would surely have reached Saili's ears.)

(I smiled)

She was my PM in the project we just finished.

The crew starting serving breakfast and we both ordered a sandwich and a coffee. After having the sandwich and the coffee I closed my eyes and immediately dozed off to sleep. I was woken up suddenly by a loud noise, I looked outside the window and I could see the clouds. I didn't know what made that loud noise but I really got scared to death. Being a frequent air traveller I always had nightmares of being in an air crash. But within seconds I realized we had just landed and it was fog outside and not clouds. My sleep had really taken over my judgement power but I was happy to be alive.

It was 9:10 a.m. The interview was scheduled to start at 11:00 a.m. I took a taxi to the hotel where I was booked. It always feels great to be back in Delhi, the place I was born and brought up. The cab driver seemed to be funny character and we chatted on our way. It was raining heavily and there was traffic all around. But the driver was no less than a F1 driver; at least his driving skills gave that impression. He dropped me off at the hotel nearly on time. I showed my vouchers at the hotel reception and checked in. I had to rush as the interview was scheduled for 11:00 a.m. and I just had half an hour to get ready. I changed into my formal wear in no time. Though I had started to feel hungry again I decided to skip my breakfast. I asked the receptionist the way to the office and she suggested I could just walk down as it was nearby. I had to borrow an umbrella from her, it felt a bit embarrassing that I didn't have my own but I had no other option. To my surprise the company office was just a ten minutes walk from the hotel. I reached precisely at 11:00 a.m.

And I thought I knew QTP!

I reported at the company's reception and the lady at the reception asked me to wait for some time. After 30 minutes or so she offered me coffee and told me that the interview would start at 12:30 noon. I silently cursed myself for skipping my breakfast. She brought a cup of coffee and few cookies as well. I was glad to see something to eat. I finished my coffee and went back to waiting for the interview call.

After some time a guy came and sat next to me. He had a file in his hand and a laptop bag. I wondered if he too had come for an interview.

Guy (Glancing sideways): Hi Nurat.

Consultant Call #2

(I wasn't sure if he actually said my name and if yes, I was trying to figure out how did he know?)

Me: Did you call me?

Guy: Yes.

Me: How do you know my name?

Guy: It's visible on your résumé

(I had my résumé in a transparent file which I had already taken out.)

Me: Oh!!!

Guy: By the way, I am Andrew.

Me: Hi, Nurat. You already know the name. Are you here for an interview?

Andrew:

(He paused for a second, smiled and then said)

Yes. What about you?

Me: *(I didn't know why he was smiling the way he did)*

Yes, I am also here for an interview.

Andrew: Okay. Which position?

Me: QTP framework designer. And you?

Andrew: Same

(And he smiled again. Though I was not scared of competition but his smile gave me a few negative vibes.)

Andrew: Did you have any rounds of interview before this?

Me: Yes, I had one telephonic interview.

Andrew: So how was it? Was it tough?

Me: No, it was pretty much related to basic stuff.

Andrew: What did they ask?

Me: Well some questions on DataTables, Actions, Object Repository and few more from here and there.

Consultant Call #2

Andrew: My friend had given one interview here earlier and they really made him nervous. I was pretty confident that he will clear the interview but he was eliminated in the first round itself.

(Well, that was not very comforting to hear, but somehow he sounded as if was trying to scare me.)

Me: I hope our nervous systems behave well today. ☺

Andrew: Yeah, I hope so too.

Me: What is your experience in IT?

Andrew: 7 years. And yours?

Me: Mine is also 7 years.

Andrew: How long have you been working with QTP?

Me: Around 6 years now.

Andrew: So you must have worked on different frameworks…

Me: Yes, Data driven, Keyword driven, Hybrid & BPT. So you can say almost all of them.

Andrew: That's good because this interview is for framework designing, so it's better to know all this stuff.

Have you worked on any open source framework?

Me: No, I haven't. You?

Andrew: Nah, neither have I. Which framework do you prefer?

Me: I usually prefer Hybrid ones.

(I was about to say more but then I just realized that he here for the same position and I better keep my answers to myself.)

Andrew: Do you know how many candidates they are going to hire?

Me: Not sure. One thing I know for sure is at least one. (I said smiling.)

Andrew: Don't be so sure. If they don't find anyone suitable for the job they don't hire.

Me: Ok.

Andrew: Are you are from Delhi?

Me: Born and brought up in Delhi but don't work here.

Andrew: So you must be used to Delhi…

Me: Of course, I am. And would love to be back here.

(Andrew's phone started ringing.)

Andrew: Hey, excuse me for a minute.

Me: Sure.

He left the scene and I waited for some more time. Finally I got the call for the first round interview. I was asked to wait inside the meeting room C2.

Personal Interview

Round 1

The meeting room was big and felt very quite. I had been waiting for the interviewer to come for over ten minutes now and the heavy silence in the room was just making my nerves jangle. I occupied my mind by wondering who the interviewer would be, what kind of interview will it turn out to be and so on...

I knew I would know the answer to all my questions in a few hours.

I could see someone walking towards the meeting room door and but I could not see the face because of the frosted strips on the meeting room glass. He entered the room and for a moment I was stunned.

Me: Hey, what are you doing here?

Andrew: Hi Nurat, I am Andrew. I am the team lead of the core framework team over here. I'm not sure if you have already been informed, but this position is for another team lead that we want to add to our team.

Me: *(I started replaying in my mind everything I had told him at the reception. It was just a generic conversation so I thought to myself that it shouldn't change the course of the interview. It also became clear to me why he was smiling earlier when he said he was also here for an interview. He was here for my interview only.)*

Yes, the consultant had informed me that your company is working on some new framework product.

Andrew: Since this requirement is for core design we are looking for QTP experts. Your feedback from the telecom interview looked good to us and I hope you won't disappoint us.

Me: Hopefully, I won't.

(The comment made me feel as if I am in military training or something....Aye Aye SIR!!!...)

Andrew: Tell me something about yourself?

Me: I am Nurat. I have done my B.E. in Computer Science from TISN College, Delhi. I have been working with Sysfokat since June 2004. During the initial years of my career I was working in manual testing and after that I have been working in QTP Automation. I worked in various projects and have created and maintained different types of frameworks as well.

Andrew: Did you prepare for this Interview?

Me: Yes, I never go unprepared. If I know about an event in advance I will always take some time to prepare well for it.

Andrew: What if you fail today?

Me: I know that an opportunity here could take my career to new heights and it would also give me chance to demonstrate my skills.

But, my being in this company will not define success or failure for me. Wherever I am, I would give my 100%. Also, success or failure is a part and parcel of life. You learn from failures and move on to new challenges with even stronger determination. So I would definitely be bit disappointed if I fail but I would take it in my stride and move on.

Andrew: How do you differentiate yourself from other candidates?

Me: The role here is for a framework designer or rather for the new framework product that your company is developing. I bring experience in developing frameworks from another company where I have done this many times before also.

In today's world one may not know about everything because of the vast scope of any given domain. But, if you know how to find what is needed and grasp it fast then you can do anything and I consider myself up to the mark at that.

Many automation candidates limit themselves to QTP or just any tool but I always had a keen interest in learning new technologies as well. I have worked a bit on C# as well though my core work is just VBScript in QTP. So I can go out of the box whenever needed.

Andrew: Why did you choose Test Automation as your career path?

Me: (That's an interesting one....I smiled...I never chose Automation, it just chose me and I stayed :D)

I always wanted to be a developer when I started out in the IT industry. But my first project brought me into QA and I continued with the same for another 2 years. After being in Automation for more than a year, I realized that my development skills and interest in the same could give me an edge in my career and I decided to strengthen my practical knowledge on QTP. From that time onwards, there has been no looking back and I still am working to enhance and upgrade my skills in Automation.

Andrew: Why do you want to leave your current company?

Me: I have been working with them for over seven years now and I have learnt a lot over there. I know their processes well, but now I feel I should move on to a new environment to expand my knowledge. This would mean new challenges, new processes to learn. I would be able to apply what I have learnt previously and at the same time learn new things as well.

Andrew: What different versions of QTP have you worked on?

Me: I started working with QTP 8.2 and after that I have worked on QTP 9.2, QTP 9.5, QTP 10. I am currently exploring QTP 11 as well.

Andrew: Do you know the release dates of these versions?

(Hell NO!!!...Am I a calendar or what??)

Me (Nervously): No, I am not very good at remembering dates. I have a hard time remembering my friends' birthdays, so this is very difficult for me. The only one I can say about is QTP 10 which was released somewhere in Feb 2009 if I am not wrong. And QTP 11 got released in Sep this year (2010).

Andrew: Ok, can you highlight the changes across these versions?

Me: I will try to highlight the major changes:

- QTP 9.x introduced debugging of libraries
- QTP 9.x restructured lot of UI menus from the previous QTP 8.x
- QTP 9.x changed the way shared ORs were used. Before QTP 9.x, we could only use 1 SOR for a test. But now we can associate multiple SOR to the Actions within the test.
- Print log was added with QTP 9.x
- The DotNetFactory object was introduced with version 9.1
- RepositoriesCollection to load repositories at run-time was added in QTP 9.2
- QTP 10 introducedIntelliSense for COMobjects
- System counters to monitor performance of an application was added in QTP 10
- QTP 10 added the option in ReportEvent to specify a image file
- QTP 10 added option to create PDF from Test Result Viewer
- Support for Web 2.0 applications was added with QTP 10.0

The new features that have been added in QTP 11 are

- Support for '.Object' property for FireFox browser
- Recording on FireFox browser
- Identify web objects using XPATH
- Web Extensibility can now be used on all supported browser
- JavaScript can be loaded and executed inside the browser
- Load function libraries at run-time and debug them. It was earlier not possible in QTP 10
- Enhanced object spy, objects can be directly added from the Object Spy itself and we can also copy object properties to the clipboard
- Exporting/importing checkpoints also with object repository is now possible. Earlier version only supported objects
- Dual monitor support. Previous versions had limitation and we had to keep the application on the Primary monitor for it to work fine
- Running scripts with remote desktop minimized. This was not possible in earlier versions
- Visual object identification. This features allows us to identify an object with reference to others objects
- New result viewer UI which has some additional graphs as well
- Log tracking for applications using Log4Net or Log4Java frameworks

These are the main core changes that happened that I feel are worth mentioning. Rest of the changes were small and didn't make a huge impact.

Andrew: Good. So what technologies have you worked on with respect to QTP?

Me: I have pre-dominantly worked on Web. But I have some experience with Siebel, Windows and Java as well.

Andrew: What are the differences between using Functions and Actions?

Me: Function is a VBscript feature which allows us to group code while Action is a QTP feature. Actions have an associated Local Object Repository, Local DataTable and can also have a Shared Object Repository while Functions cannot be associated with these. A Function can be defined inside an Action while an Action cannot be defined inside a Function. Similarly, a Function can also be stored inside or outside a QTP's Test. An Action is a component within the Test.

Moreover, a Function can be loaded into and referred from the memory whereas an Action cannot. It is interpreted and executed at run-time. This generally results in Functions having better performance as compared to Actions.

Unlike Functions, Actions can return multiple values through Output parameters. Functions return a single value, but their usage can be manipulated through the use of ByRef, Collections and Public variables which is an indirect way of returning multiple values.

Andrew: Can I call a Function in one Action from another Action?

Me: No. Every Action in QTP runs within its own namespace and the only namespace it has access to, is the global library. So, one Action cannot call an Function from another Action.

Andrew: Which is better to use: Actions or Functions?

Me: Well there are few pros and cons in both approaches but I see more downsides in using Actions.

Some key points when using Actions are

- Actions allow us to have optional parameters.

- Action Parameters cannot accept complex values like Array or object.

- A blank Action with no object and no code occupies 150+KB of space, so using too many Actions increases the size of scripts.

- Re-usable Actions, when called in Test are read-only. So, if there is an issue in running the Test and the Action's code needs to be updated for fixing it, then we need to close the Test and re-open it containing the Re-usable Action. After making the changes to the Action code we need to open the test again for execution and in case it fails we need to repeat the procedure for fixing the code. This is a huge maintenance issue while using Actions.

- Actions create a separate section in the Test results summary making it easier to differentiate functionality.

Compared to Actions, Functions have a few advantages:

- Functions don't take much space as they don't have associated Local OR or DataTable.

- Functions can easily be overridden by redefining them.

- Functions can be edited easily as compared to external Actions. This makes maintenance much more easier.

- Functions parameters can take any possible value supported by VBScript.

So, I personally prefer Functions as maintenance becomes a nightmare when we have too many Actions.

Andrew: What are the default Add-ins that come with QTP?

Me: QTP 9.5 and higher have all Add-ins packed with the installation. During the installation, we can customize which particular Add-in needs to be installed. The default installation has Web, ActiveX and VB selected.

And I thought I knew QTP!

Andrew: How do we load an Action at Run-time?

Me: QTP provides a LoadAndRunAction method that can be used to do the same. But this is only available in QTP 10 or higher.

Andrew: How would you do that in versions lower than QTP 10?

Me: There is no way to do it in QTP 9.5 or lower. As a workaround, we can read the code of the Action from the Test and execute that at run-time, but we would be looking at lots of issues like not having Local OR, DataTable, Associated Shared OR, and parameters. Overcoming these is nearly impossible as QTP uses binary formats for storing such information.

 Note: Dynamic Action call is not supported on these versions but there exists a two-part article which demonstrates how to achieve this in QTP 9.2 and lower

KnowledgeInbox.com/?s=dynamic+action+call

Andrew: I have 3 Actions in my script – Login, PlaceOrders and Logout. There are 4 iterations in my Global DataTable and I want the Login and Logout to be performed only once, Login for before the first iteration and Logout only the last iterations. How will I achieve that?

Me: If we had the case of Login being only called once, I would have used a Global variable in one of the associated library files:

```
Dim bLoginDone: bLoginDone = False
```

And in my login Action, I can check for this flag and exit the Action in case it is already executed

```
If bLoginDone Then ExitAction "Login Already executed"
bLoginDone = True
```

Personal Interview–Round 1

But, since we want to do this for last Logout Action at the end only then I don't think this approach would be of much use. So, what we can do is to go to the Keyword view and delete the call to Login and Logout action and call only the PlaceOrders action. Now, in order to execute the Login and Logout action I will take help of class Initialize and Terminate event

```
Class ActionLoader
   Sub Class_Initialize()
     RunAction("Login")
   End Sub
   Sub Class_Terminate()
     RunAction("Logout")
   End Sub
End Class

Set oActionLoader = New ActionLoader
```

The above code would make sure that when the test is initiated, the Login Action gets executed. And when the Test is about to end, the Logout Action is called.

Andrew: How do you delete an Action call?

Me: If we remove the RunAction statement from the Expert View, it will remove the Action from TestFlow but the information of the Action would still remain in the Test. If we want to completely delete the Action then we need to navigate to the Keyword View, right click on the Action call and select delete from the list. We would be given an option to delete the call as well as the Action information from the Test.

Andrew: Can we hide Keyword view in QTP?

Me: Yes we can from QTP setting. But this is only in case of QTP 11.

And I thought I knew QTP!

Andrew: What is the difference between a Function and a Sub-routine?

Me: A Function can return a value while a Sub-routine cannot return a value.

Andrew: Can you give me a small example on how would you choose implementing something as Function or a Sub-routine.

Me: Yes, consider that we have to create two methods: one to check for file existence and one for deleting the file. Now, since the result of file existence check would be required to return a Boolean. In other words, we need to derive 'True' if the file exists and 'False' if it doesn't. In this case, I would use a Function as it is returning a value. In case of DeleteFile, I want the file to be deleted. This case is a little different from the one I just mentioned because if we fail to delete the file, I would raise an exception rather than returning a False value indicating that the operation failed. So, in this case I would use a Sub-routine.

Andrew: How can I return multiple values from a Function?

Me: A Function can only return a single value. But in case we are interested in extracting more data, then we can use ByRef parameters and return the value using the same. Since Collections like Arrays and 'Scripting.Dictionary' can hold multiple elements, we can also pass multiple values as elements of these collections. Another way to return multiple data is through the use of class. I can create a Class and add multiple properties to it. Then, from the function I can return the object of the Class.

Andrew: Can you give me sample code for this class approach?

Me: Yes. Something on below lines:

```
Class MultiData
   Dim Info1, Info2, Info3
End Class

Function GetInfo()
   Dim oInfo
```

```
    oInfo = New MultiData
    oInfo.Info1 = "Info1"
    oInfo.Info2 = "Info2"
    oInfo.Info2 = "Info3"
    GetInfo = oInfo
End Function

Dim oInfo
Set oInfo = GetInfo

Msgbox oInfo.Info1
```

Andrew: And how about the dictionary approach?

Me: One way is to return the dictionary from the Function and the other is to pass the dictionary as parameter and then load it

```
Function ReturnDictionary()
   Dim oDict
   Set oDict = CreateObject("Scripting.Dictionary")
   oDict("Info1") = "Information1"
   oDict("Info2") = "Information2"
   oDict("Info3") = "Information3"

   Set ReturnDictionary = oDict
End Function
```

Another way is to pass the dictionary

```
Dim oMyDic
Set oMyDic = CreateObject("Scripting.Dictionary")
Call LoadDictionary(oMyDic)
```

And I thought I knew QTP!

```
Function LoadDictionary(oDict)
   oDict("Info1") = "Information1"
   oDict("Info2") = "Information2"
   oDict("Info3") = "Information3"
End Function
```

Andrew: What are the two ways of passing parameters to Functions?

Me: One is ByVal and the other one is ByRef.

Andrew: What is the difference between the two?

Me: If we pass a variable to a ByVal parameter, VBScript copies that value to the Function parameter and any changes we make to the variable in the Function don't reflect in the variable passed. If we pass a variable to a ByRef parameter, then the parameter contains the reference to the original variable used and any changes made in the Function would also be reflected to the original variable.

Andrew: In case I omit specifying the parameter as ByVal or ByRef then what would happen?

Me: By default the parameters are passed as ByRef

Andrew: Ok. I have two Functions, one which takes a ByVal parameter and one which takes a ByRef parameter. I pass an array to both of them and inside the Function I change one of the elements of the Array. In which case would the changes made inside the Function be visible?

Me: I have never tried this out ever but I see no reason it should not work the same way it does with other variables data types. So the ByVal Function changes should not impact the array but with ByRef, it should.

 Note: Nurat's understanding is correct regarding above

Personal Interview–Round 1

Andrew: How do you implement a Continue statement in a loop?

Me: Well, neither QTP nor VBScript provide any Continue statements.

Andrew: Can you think of some workaround?

Me: Let me think

(I need something that goes to the Next statement from in between. This can only happen if I use an Exit Statement, but I can't use Exit For as it would exit the main loop)

I think it would be possible using an extra For loop inside the existing loop. E.g. If we want to continue when the loop counter is divisible by 3, then I can implement the code in the following fashion

```
For i = 0 to 20
   For j = 0 to 0
      If (i Mod 3) = 0 Then Exit For
      'My original code
   Next
Next
```

Andrew: This looks fine but what if inside my code I have a condition where I had used Exit For and it was supposed to exit the main loop. Wouldn't your approach break my code?

Me: *(Ahhh....will he ever be satisfied with my answers :-( )*

Well in that case I can use a Do loop instead of a For loop and the new code would be as below:

```
For i = 0 to 20
   Do
      If (i Mod 3) = 0 Then Exit Do
      'My original code
```

Lalwani

And I thought I knew QTP!

```
  Loop While False
Next
```

This would work without impacting your existing code

Andrew: Tell me what methods and properties are supported by the Reporter object and what is their use?

Me: The reporter object supports ReportEvent, RunStatus, ReportPath, ReportNote and Filter.

ReportEvent is for adding information to the test results

RunStatus is for getting the current status of the test at run-time, whether it has failed or passed

ReportPath gives the folder location where the results are being stored

ReportNote allows adding notes to the report. This method has been added in QTP 11

Filter is used to disable or enable reporting of events to the Test's report

Andrew: How do you check if the Test has passed or failed?

Me: We can use the RunStatus and check its value. A micFail value would indicate that some step within the test had failed.

Andrew: If I run an Action for 5 iterations, can I check their status using RunStatus?

Me: No. Once there is a failure in the test the RunStatus would be set as micFail. It would not change even if further steps pass.

Andrew: Can't we reset this RunStatus?

Me: No it is a read-only value

Personal Interview–Round 1

Andrew: How would you check the status of iterations then?

Me: There is no direct way to do this. Failures can be reported in different ways – Script errors, Checkpoint failures and user reported errors. We can use Recovery scenarios with a Function call to set a failure flag. This flag can be reset before executing the iteration of the Action and we can read the flag value at the end of iteration to see if the Action passed or not. But this approach would only count for errors that can trigger the Recovery Scenario. For custom errors reported through Reporter.ReportEvent, we will have to add the code to set the failure flag.

 Note: This approach may not be as easy as it seems. To tackle this we need to capture failures in various ways:

1 - *Introducing a flag to check for error*

This can be done by using environment variable

```
Environment("Action_Passed") = True
```

The code needs to be added to the top of the Action so that the flag is set before performing any steps

2 - *Setting the flag in case an error occurs. This can be done in 2 ways:*

Using On Error Resume Next

We can move all the code of the Action into a Function and then call this Function as below

```
Function ActionCode()
End Function

On Error Resume Next

Call ActionCode()
```

Lalwani

And I thought I knew QTP!

```
        If Err.Number <> 0 Then
            'An error had occurred
            Environment("Action_Passed") = False
        End If
```

This can cover up for most of the errors but not for CheckPoints and custom reported failures. For catching Checkpoint failures we need to override the Check method of all objects for which we have used the Checkpoints

```
Function NewCheck(ByVal Obj, ByVal CheckObj)
    Dim bCheckStatus
    bCheckStatus = Obj.Check(CheckObj)
    If Not bCheckStatus Then
    Environmen("Action_Passed") = False
    End If
End Function

RegisterUserFunc "WinEdit", "Check", "NewCheck"
```

For capturing custom reported failures we need to make sure that instead of calling Reporter.ReportEvent method directly we call an additional wrapper Function and then set the failures:

```
Function ReportEvent(ByVal Status, ByVal StepName, ByVal Description)
    Reporter.ReportEvent(Status, StepName, Description)
    If Status = micFail Then
        Environmen("Action_Passed") = False
    End If
End Function
```

The approach above sets only one flag with name as Action_Passed. But to support multiple Actions we can generate the environment variable name at run-time using the current Actions name

```
envName = Environment("ActionName") & "_Passed"
Environment(envName) = True
```

Andrew: What values can the status take in ReportEvent?

Me: micPass, micFail, micWarning, micDone and micInfo

Andrew: How do you import a DataTable in QTP?

Me: We can 'DataTable.Import' in case we want to import the whole file. If we need any specific sheet then we can use 'DataTable.ImportSheet'

Andrew: Any other way you can import Data into DataTable?

Me: Another way is to read the Excel file using Excel COM API and then populate the data cell by cell and row by row. But this would give very poor performance when compared to the DataTable.Import method.

Andrew: I want to add an object to the OR but my OR is full what can I do?

Me: (Full!!!!!!!! I have never heard that in my life)

I have worked with QTP for over 5 years and till now I have not seen any error message which says QTP's OR is full. So, in my personal experience, this would never happen. But, in case for some weird reason it does happen then I would try few options like removing unwanted objects, or creating new shared object repository. Also, at the same time I will open a ticket with HP to explain to me why there is a maximum size and even if there is a limit then why is it not documented.

Andrew: How can we add objects to Object Repository at run-time?

And I thought I knew QTP!

Me: We can only add ObjectRepositories at run-time and not objects to Object Repository at run-time. To work with objects at run-time, we can use Descriptive Programming (DP).

Andrew: We have a Test script that was created using objects from a Shared Object Repository. Now, we want all the objects to be moved into the Local Object Repository and remove the dependency on the Shared OR. How can we do this?

Me: Well, QTP doesn't provide any way of importing objects from a Shared Object Repository to Local Object Repository. But the Object Repository allows right clicking an object and using the option 'Copy to Local'. This can be used on multiple objects at the same time but only for objects with common parents. So if the Shared OR is huge, it would be a cumbersome task to copy everything into the Local OR.

Andrew: What is an Object Repository parameter?

Me: To parameterize object property values in Object Repository, we can use Object Repository parameters. Shared Object Repositories do not have access to Environment variables or DataTable. When we add such a Repository to our Test, the parameters can be mapped to an Environment variable to a DataTable parameter.

Andrew: Can we update the Object Repository parameter at run-time also?

Me: Yes, it can be done using the Repository object.

```
Repository("ParamName") = "Value"
```

Andrew: How can I load an Object Repository at run-time?

Me: We can use 'RepositoriesCollection.Add' to add Object Repositories at run-time in QTP. But this is only available from QTP 9.2 onwards.

Andrew: How do you merge two shared Object Repository into one?

Me: The Object Repository manager tool has an option to launch the Object Repository merge tool. This merge tool takes the two input files that need to merged and then allows saving the merged file into a new file.

Andrew: What happens in case there are any conflicts during the merge process?

Me: There are 3 types of conflicts

- *Similar description conflict – If two objects have the same logical name and hierarchy but one object has additional properties in comparison to the other one. In this case we can choose the keep the less generic or more generic description.*

- *Same name different description conflict – If two objects have the same logical name and hierarchy but different descriptions. They can use different properties or have same property with different values. In this case we choose the object from one of the Repository or keep both of them.*

- *Same description different names conflict – If two objects have the same parent objects and same description but different logical names. In this case, we can choose to keep one of the objects from the primary or the secondary Repository. We cannot keep both the objects in this case.*

Andrew: How do you encrypt passwords in QTP?

Me: To encrypt a password at design time we can use the Password encoder tool. To do it at run-time we can use the Encrypt method of the Crypt utility object

Andrew: How do you decrypt this password?

Me: There is no method provided by QTP for this. But this encrypted password can be used with SetSecure method on a TextBox and retrieving the value of the TextBox will reveal the password.

Andrew: How do you perform Cross platform testing with QTP?

Me: QTP only supports the Windows platform. If by Cross platform testing, you mean various versions of Windows, then that can be done by just running the scripts on different machines with different Windows OS versions. If you mean Cross platform as other OS's like Linux, UNIX etc, then QTP is not the right tool for that.

Caller: What does 'mic' in micClass stands for?

Me: I don't know if it is documented anywhere by HP but I guess a reasonable full form could be 'Mercury Interactive Class' as QTP was earlier a Mercury Interactive product.

Actually, we have constants like micPass, micFail as well, so I think I would rather say the closest I can get is 'Mercury Interactive Constant'. But only HP or Mercury can confirm if they had the same intention when using mic in these constants.

Andrew: What is Descriptive Programming?

Me: Descriptive Programming is a feature of QTP which allows working on an object at run-time by describing their properties. This feature allows automation developers to decide which properties they would like to choose to identify the object instead of having QTP to do so with the help of its recognition process and storing properties in Object Repository. In other words, for the same Class or Type of object, users can select different properties which are not supposed by Object Repository.

Andrew: Can you please elaborate on your last statement?

Me: By default, we can only specify QTP to use a set of properties with a certain type of Test Object. For example, if we select 'Name' and 'HTML Tag' for a TextBox, QTP will always use these properties for the TextBox object. However, with Descriptive Programming, automation developers have the flexibility to select the value depending on what works well. In other words, instead of always using 'Name' and 'HTML Tag' for the TextBox, automation developers can decide to use 'HTML ID' only for one of the TextBoxes and only 'Name' for another.

In simple terms then, it's more flexible but does require manual selection of properties so there can be a design-time overhead creating scripts on an application with hundreds of objects.

Andrew: What are the different types of Descriptive Programming?

Me: DP is just DP. There are no types available.

Andrew: Haven't you heard about string based and object based DP?

Me: Oh that, I would consider that more as styles of DP rather than types. There are two styles

- In String based DP, we use string parameters and describe the objects in this:

```
Set oBrowser = Browser("micclass:=Browser")
```

- In Object based DP, we create a description object and add properties to it and then we use the object as the parameter

```
Set oBrw = Description.Create
oBrw("micclass").Value = "Browser"
Browser(oBrw).Close
```

Andrew: Which approach is better to use? Object based or DP based?

Me: Both the approaches have their pros and cons. Strings takes less memory as compared to objects. So an approach totally based on the Description object might take too much of memory whereas using a string description can increase the number of lines of code; though this can be avoided through the use of variables. Another advantage of object DP is that we can push more than one property in the description. So if in future, additional properties are required for identification of the object then no change to the code is required but in string DP an additional parameter would be required.

So, I personally favor a hybrid approach where we define all the descriptions as string and if they have more than one property to be used then I use object based DP for those.

Andrew: I have a QTP Test Object at run-time and I have identified using an Index:=1, now I want to retrieve the Location property of the same. What would I use in this case – GetROProperty or GetTOProperty?

Me: We cannot use GetROProperty or GetTOProperty to get the Location of object. Ordinal identifiers can only be used in identification of an object and cannot be retrieved at run-time.

Andrew: So you are saying it is impossible for me to tell the location of this object?

Me: It's not impossible and there is a possible workaround. If this is a Windows based application then each object will have a handle that can be read using GetROProperty('hwnd'). So, I will have to get the handle of object we identified using index:=1 and then I will have to get the object using location:=i, where I will be keep on incrementing the value of 'i' till I get the desired handle. To illustrate, please allow me to show you the following code for a WinEdit:

```
'Retrieve the Handle of the target WinEdit
iHWND = Window("").WinEdit("").GetROProperty("hwnd")

Set oDesc = Description.Create

oDesc("micclass").Value = "WinEdit"

Set oParent = Window("").ChildObjects(oDesc)

For i = 0 to oParent.Count - 1
   If Window("").WinEdit("location:=" & i).GetROProperty("hwnd") = iHWND Then

      MsgBox "Location Ordinal Identifier = " & i

      Exit For
   End If
Next
```

Andrew: I have launched a single browser and I am checking existence of 3rd browser using the below code

```
bExist = Browser("index:=2").Exist(0)
```

Will this give me an error or will it give me True/False?

Me: This would return true

Andrew: Why?

Me: Ordinal identifiers are only used by QTP when the other properties result in multiple matches. Since, in this case there is only one browser open, QTP will not use the index and will return true.

Andrew: But index:=2 is the only property I have specified then how come it doesn't use the index property and there is no other property to refer to as well?

Me: Every object has a default property which is micclass and it is assigned based on the type of object we use. So Browser("index:=2") is equivalent to Browser("micclass:=Browser", "index:=2"), that is why QTP doesn't use the index. Because at run-time, the Index is not used, QTP replaces it with the default Class Type which in our case is the Browser. However, it is to be noted that Browser("micclass:=Browser") is not equivalent to Browser() or Browser(""). If you provide a Null description, QTP will throw an error.

Andrew: How can I fix this then?

Me: We can use an invalid index of -1 and then check the existence of the object. -1 is an invalid Index value because it is zero-based, numeric values should begin at 0. They cannot begin with a negative value so if the object exists with index as -1 then it would mean that the index was not used during recognition. As a result of this, we can deduce that there is only one object available.

If the object doesn't exist with -1 then the object may not exist at all or it may have multiple instances available. We can test this using the Exist statement on the object. The code would be something like below:

```
bInvalidExist = Browser("index:=-1").Exist(0)

If bInvalidExist Then
   Msgbox "Only 1 browser exists"
Else
   bExist = Browser("index:=2").Exist(0)
   Msgbox "Browser Exists - " & bExist
End If
```

Andrew: How do you get the count of browsers open?

Me: We can use the Desktop object get all the ChildObjects matching the Browser and then get the count

```
Set oDescBrowser = Description.Create
oDescBrowser("micclass").value = "Browser"

Set allBrowsers = Desktop.ChildObjects(oDescBrowser)

Msgbox allBrowsers.Count
```

Andrew: I have a text box on my web application and I use the following code in DP to identify the same

```
Browser().Page().WebEdit("name:=txt$content1$test",    "index:=1").Set "Test"
```

Personal Interview–Round 1

Now the there is only one textbox with the name 'txt$content1$test', will the above code work?

Me: The code won't work but the reason behind that would not be the index. The problem is with the name. The name of the text box includes some regular expression characters and we didn't escape them. There are two ways to fix it:

One is to escape all the pattern character

```
Browser().Page().WebEdit("name:=txt\$content1\$test", "index:=1").Set "Test"
```

Another way it to use the Description object

```
Dim oDesc
Set oDesc = Description.Create

oDesc("name").Value = "txt\$content1\$test"
oDesc("name").RegularExpression = False

Browser().Page().WebEdit(oDesc).Set "Test"
```

Andrew: How do you close all open browsers?

Me: We can use

```
SystemUtil.CloseProcessByName "iexplore.exe"
```

Andrew: Any other ways?

Me: Yes we can enumerate all the browser objects and close them one by one:

```
Set oDescBrowser = Description.Create
oDescBrowser("micclass").value = "Browser"
```

Lalwani

And I thought I knew QTP!

```
Set allBrowsers = Desktop.ChildObjects(oDescBrowser)

For i = 0 to allBrowsers.Count - 1
   allBrowsers(i).Close
Next
```

A few other ways of doing the same are through a browser's title using 'SystemUtil.CloseProcessByWindTitle', its Process ID using 'SystemUtil.CloseProcessByID', using WMI etc

 Note: To close process using WMI we can use below code

```
'Name/IP of the computer
sComp = "."
'Get the WMI object
Set WMI = GetObject("winmgmts:\\" & sComp & "\root\cimv2")

'Get collection of processes for with name iexplore.exe
Set allIE = WMI.ExecQuery("Select * from Win32_Process Where Name = 'iexplore.exe'")

'Loop through each process and terminate it
For Each IE in allIE
    IE.Terminate()
Next
```

Andrew: What is Object Spy and what information does it provide?

Me: Object Spy is a QTP tool which helps us spy on objects in the Application and see their methods and properties. It has two radio buttons: one for run-time object properties and the other for Test or Native object properties. There are also two tabs to show methods and properties supported by the different types of objects.

Andrew: What is the difference between run-time object properties and Test Object properties?

Me: Test Object or rather Native object properties are the properties which are used for identifying an object. They are the properties used in OR or DP. Run-time properties are the actual properties which the identified object poses.

Caller: Can you give me an example of this?

Me: Consider we identify a Logout link with text as Logout.*. Now "Logout.*" here represents the Test Object property and if we use GetROProperty to get the text of the link it may come as 'Logout Nurat' which is the actual text of the link. This is run-time property.

Caller: But when we select the 'Run-time Object' option in the Object Spy, it shows various properties which GetROProperty actually doesn't support. Why?

Me: It is a confusion that was created by QTP 9.5 and lower versions. In QTP 10 HP realized this and renamed the option as Native instead of Run-time. Native properties are those properties which are internally supported by underlying technology of the AUT.

Andrew: How do we access the Native object properties?

Me: We need to use .Object property of the QTP Test Object and then we can use any of the methods and properties supported by the object. In case of web objects few of the native properties can also be read using the GetROProperty method with 'attribute/<attributename>' format

Andrew: How do we change the value of a Run-time object property? Can we use SetTOProperty for the same?

Me: No, we cannot directly change any of the Test Object properties. SetTOProperty is used to change the value of property used for object identification. To change a Run-time Object property, we need to use the methods supported by QTP to change them. E.g. to change 'value' property of a WebEdit we need to use the Set method which internally updates the same.

And I thought I knew QTP!

Andrew: Any other way of changing these Run-time object properties?

Me: Yes, we can use the underlying native object properties. However, this option is not supported by all Add-ins.

Andrew: I have recorded a script in Firefox and I replay it on Internet Explorer. Will it work?

Me: Recording of scripts in Firefox can only be done in QTP 11. But this feature is not available in QTP 10 or lower, so the script needs to be record in IE in such cases. As far as the compatibility of the recorded code depends on few things:

- The Firefox version being used should be supported by the QTP version
- If there are IE dialogs that got recorded like Password window, security alert, certificate errors etc. then, these would error out in Firefox as it uses totally different dialogs
- If we have accessed HTML DOM of the objects in our script then it won't work in Firefox. However, because we have recorded the script, this won't happen in our case.
- Then there are few occasional issues as well where the code that is supposed to work the same way in Firefox as well doesn't work. The ones that I have noticed is usually with ChildObjects where the description used returns the correct object count in IE returns 0 elements in FF.

Andrew: Why does QTP not support DOM objects in Firefox?

Me: (Because I don't work for HP J....)

QTP 11 now does support DOM on Firefox as well. But the DOM of Firefox is a bit different from the the IE DOM.

Andrew: How do you check if a page has completely loaded in Firefox?

Me: I will just use 'Browser.Sync'

Andrew: Which browsers are supported by QTP?

Me: QTP supports IE, Firefox and Netscape.

(I Hope he is not going to ask why Opera is not supported...)

Andrew: I start recording on IE and instead of recording on IE window as a Browser object, it records it as a Window object. What could be the possible reasons you can think of that could be causing this issue?

Me: Well there can be few possible reasons I am aware of

- *The web Add-in may not be loaded*
- *The browser version may not be supported by QTP*
- *Third party browser extensions might be disabled in IE settings*
- *QTP's BHOManager add-on may be disabled in IE*
- *QTP's installation might be corrupt*
- *QTP is launched after the target browser object*

(mmm...I am missing few points may be...)

These are the key issues I can remember.

Just one more thing to add to here is the UAC settings on Windows Vista. If UAC is enabled then it can also cause issues with IE.

Andrew: You are given a script which executes unusually slow. What are the possible reasons you can think of which could be making the script perform slow?

Me: Well there could be many reasons behind this

- There might be too many unwanted wait statements with high timeouts.

- If Smart Identification is enabled and it gets used, that can also slow down the script.

- Checking objects with same timeout on the same page. E.g. Consider a screen with 5 objects and I use a timeout of 20 seconds on each. Due to a server error, the screen does not load correctly and the very first object my Exist statement checks that it is not found. This behaviour will cause my script to wait for 100 seconds. Since we are giving 20 sec for wait in initial object we are assuming that the screen should appear in 20 seconds. Therefore, for the other 4 objects, we can assume that the screen is already loaded and we can change the timeout to 0 seconds from 20 seconds.

- If the ChildObjects method has been used several times in our test to match a small number of objects in a very heavy HTML/JavaScript page, we may experience delays in execution time.

- By default, QTP uses the object synchronization timeout which is defaulted to 20 seconds. It also uses a WebTimeout which is defaulted to 60 seconds. We can lower these settings according to our application's needs and optimize the wait times.

- Loops are not exited from, once the condition has been met. For example: consider you are checking attributes of a List object in a Web application. Now, you would like to only check for the tagName property. Once the tagName property is found from the attributes collection, the loop continues instead of being exited causing QTP to use extra time to complete the entire loop. In short, always Exit loops once conditions are met.

These are all the points I can think of as of now.

Andrew: Any other thing related to QTP's settings?

Me: (Did I miss anything??....In file settings I have Object Timeout, WebTimeout, Smart Identification. So that is already covered. In tools->options I have run mode...May be he is looking for Run Mode...)

Yes, I missed QTP's Run Mode setting. If it is set to Normal mode with some per step delay then the script takes more time. In the Fast mode, the script executes without these delays, but the execution marker is not shown.

Andrew: What is the difference between options given in File->Settings... and Tools->Options...?

Me: Whatever we change in File->Settings... is saved and associated with the script. If we open the script in another machine these settings would be the same. Tools->Options... settings are user or machine specific. These settings are not associated with the script.

Andrew: What are synchronization points?

Me: Synchronization points help QTP to wait for the application to achieve certain state where next operation can be executed. Without synchronization points QTP may act too fast on the application causing the scripts to fail. For example, in a web application, once we click 'Add to Cart', it may take a while before the items are successfully added to the cart. A synchronization point can be added here to counter the delay application experiences before items are added to the cart feature. This ensures a smooth running of the script and QTP does not error out because of an 'Object Not Found' error.

Andrew: How do you use synchronization in scripts?

Me: It all depends on what kind of situation we are in

- If we want to wait for the page to load in a Web application then I will use Browser.Sync
- If we want some property on the web page to change then I would use WaitProperty method
- We can also use object existence or non-existence as a check if that is possible using the Exist method

Andrew: What is the difference between 'Browser.Sync' and 'Page.Sync'?

Me: QTP doesn't document any difference between these two Sync methods. But I have experienced that if we use 'Browser.Sync' then it is more helpful in application with Frames. But there is no documentation support for this.

Andrew: What would you prefer? A high performance script which has low wait times but can fail occasionally or a script with more wait times and higher reliability?

Me: It is a tough choice to make and we need to make sure there is balance. When I started my career in Automation, I always used less wait times and that meant my scripts completed very fast. Though there used to occasional failures, but the impact wasn't much as the scripts ran very fast. But when we implemented this approach in big Automation projects with over 1000+ scripts we realized that a small failure of script due to low wait times caused us to spend a lot of hours on result analysis. So, we felt that is important to make scripts more reliable by compromising the speed. And we were able to reduce a lot of Analysis time in later releases. Now I usually prefer scripts to have enough time before failing.

Andrew: Have you worked with automation of Office applications like Excel?

Me: (Smiling) Yes I have.

Andrew: Ok, I have an Excel sheet that is already open. How can I get access to the worksheet object of this Excel?

Me: We can use GetObject and pass the Path of the Excel file. That should give us the access to the WorkBook object:

```
Set xlsWorkbook = GetObject("C:\Test\MyXLS.xls")
```

Andrew: What if I don't know the path?

Me: Then instead of this I would use the class of the Excel application:

```
Set xlsApp = GetObject(, "Excel.Application")
```

```
Set xlsWorkBook = xlsApp.ActiveWorkbook
```

Andrew: How does this GetObject work?

Me: I don't know the complete internals of this but I know it at a very high level. Applications can register themselves in something called ROT. It is Running Object Table. Once the application registers itself in ROT then any COM application can use GetObject method to get the object using a specified text that may be defined by the application.

Andrew: How do CreateObject and GetObject differ?

Me: CreateObject is to create new instance of an object while GetObject is used to bind to an existing object instance.

Andrew: What happens when you have two Excel instances open and you use GetObject?

Me: You will get access to any one of them and as per my knowledge goes, you have no control over which instance is provided to you.

Note: GetObject returns the instance which was launched first and not just any random instance.

Andrew: I want to launch IE using CreateObject, how would I do it?

Me: We would instantiate the object using its (classic) COM class name and set the visible property to True

```
Set oIE = CreateObject("InternetExplorer.Application")
oIE.Visible = True
```

Andrew: Ok, If I add a statement 'Set oIE = Nothing' in the above code and run it. What happens to the current IE session when my script ends?

Me: Nothing, you will still see the IE window opened. Only the reference to the object will be released.

Andrew: Doesn't it defy the rules of object destruction? When I have one object and only reference to it, the object should be destroyed on setting the last reference to Nothing

Me: COM objects can run in two modes: in-process or out-process. The in-process COM objects load itself into the process which creates the object while the out-process objects run outside the process domain. All in-process objects get destroyed when their reference count becomes 0 or the when the application closes. But an out-process COM object has control over this destruction and can live even after the creating process terminates. Excel, Outlook, Word all such applications run in out-process mode and doesn't get destroyed when the creator application terminates. To close the application we need to make sure we call on the Quit method on the main application object.

Andrew: Which method do you think is better to use, CreateObject or 'SystemUtil.Run'?

Me: I prefer using 'SystemUtil.Run' because, first of all CreateObject is not supported by all applications. Also, launching IE using this approach has its side effects as the new browser inherits the session information of the last browser launched. So using this approach in iterations may not work well starting at the 2nd iteration.

Andrew: Can I use CreateObject to launch Firefox as below?

```
Set oFF = CreateObject("Firefox.Application")
oFF.Visible = True
```

Me: No. Firefox is not a COM based application. CreateObject is only supported for applications which are built over the COM architecture. We can't just randomly use this for any application.

Andrew: What is COM?

Me: COM is Component Object Model. It is a standard format of creating language independent components which can be used by any language that supports COM. It is mostly used by applications which need to provide scripting or automation capabilities. Most of the office products like Office, Outlook, Excel, and Internet Explorer expose their functionalities through a set of the COM objects.

Andrew: What about DCOM?

Me: DCOM means Distributed COM. It is when COM objects are created and used across network.

Andrew: What else?

Me: I have not done DCOM programming so this is what I know.

Andrew: Then how would you know an Application is supported by CreateObject. Or rather how do you tell if an application is a COM based application or not?

Me: Most applications which are developed for COM Automation, they document their object model in the developer reference section of the help. So that is the best place to start looking for if the application supports COM or not.

Andrew: How does CreateObject internally work?

Me: It takes in the input as the ProgID which is given by the developer of the COM components. It then internally uses Win32 COM API to create the object. I am not very sure as to what these APIs are

Note: CreateObject takes the input as the ProgID of the COM component to be created.

It then internally searches for the ProgID in system registry in the HKEY_CLASSES_ROOT

If the key doesn't exist then the CreateObject method would throw an error. If it exists then the CLSID of the same is captured. Consider the below code

Set oDic = CreateObject("Scripting.Dictionary")

And I thought I knew QTP!

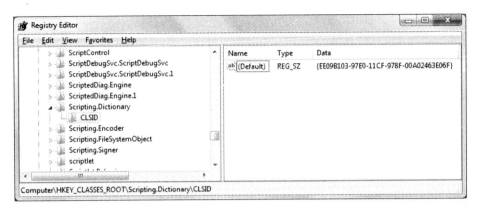

CLSID for the ProgID "Scripting.Dictionary" is '{EE09B103-97E0-11CF-978F-00A02463E06F}'. CLSID also known as Class Identifier are unique 128-bit GUID used to uniquely identify a COM class present on the system.

Once the CLSID of the class it available the CreateObject methods look for the details for the corresponding CLSID in HKEY_CLASSES_ROOT\CLSID\ key in the registry.

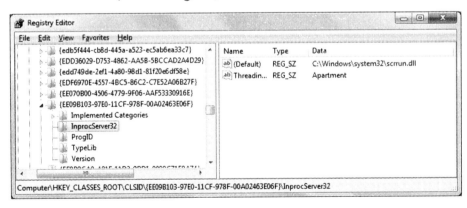

Personal Interview–Round 1

> The InprocServer32 key gives the location of the DLL or EXE file which hosts the COM server. In case of DLLs the key name is InprocServer32 and in case of EXE the key name is LocalServer32
>
> The object is then instantiated using Windows API

Andrew: How do you know which methods are supported by a COM object?

Me: There are various ways of knowing what methods a COM object supports

- One is to look in the developer reference documentation of the COM component

- We can use a tool like OLE COM Object Viewer and view the type library information of the component. This tool shows all the Interfaces and Classes supported by the component

- We can use any development tool like Office Products Macro Editor, VB6, Visual Studio .NET and add the reference to the COM component. The methods of the component can then be viewed in the object browser

> Note: OLE COM Object Viewer can be downloaded from the below link

> Microsoft.com/downloads/en/details.aspx?FamilyID=5233670d-d9b2-4cb5-aeb6-45664be858b6&displaylang=en
>
> ActiveXplorer is another tool that can be used for exploring COM components

> Aivosto.com/activexplorer.html

Andrew: How can we incorporate error handling in QTP?

Me: There are 4 different ways to add error handling:

Lalwani

- One is to use QTP's Test Settings where we can set what happens when an error occurs
- Second is to use the VBScript 'On Error Resume Next' statement
- Third option is to use recovery scenarios
- Error Handling can also be performed using Conditional statements but here, a user generally expects an error on a certain point and uses the conditional statement to tackle it.

Andrew: What options for error handling does QTP Test Settings provide?

Me: One is to raise the error which gives you an error dialog. Another option is to skip to the next step. Another option we have is to Exit the Action iteration and the last option is to stop the test run.

Andrew: How do you use On Error Resume Next to check for errors?

Me: If we use On Error Resume Next, the code moves to the next statement and VBScript updates the Err object with the details of the error. We can later use these details in code to check for specific errors or log these errors

```
On Error Resume Next
   X = 2/0
Msgbox Err.Description
```

Andrew: How do you cancel or disable 'On Error Resume Next' statement?

Me: We can use the 'On Error Goto 0' statement in the code. This disables the 'On Error Resume Next' statement.

Andrew: If I have 20 lines of code and I want that if any error occurs on any of the lines then the rest of the code should not be executed. Then what should I do?

Personal Interview–Round 1

Me: In such cases the better option would be to move all the 20 lines of code into a Function and then in code use On Error Resume Next and then call the Function. When error occurs in the Function, since there is no error handler inside the Function the code would exit out of the Function and we can then catch the error

```
On Error Resume Next
  Call MyFunction

  If err.Number then
    Msgbox "Error - " & Err.Description
  End if
On Error Goto 0
```

Andrew: Should we use On Error Resume Next in each script?

Me: No, that could point to a real poor design. One of the reasons is that, we need our code to check the error description ('Err.Description') and number ('Err.Number') at the point where the error occurs. If our check is located a few lines below or above the error, then it can become quite confusing to spot the line where the error originally occurred. This can make debugging tough. A better solution to debugging such errors would be with a Try-Catch-Finally statement which is not available with VBScript.

Another issue at times is where we use On Error Resume Next and expect an error to occur which needs to be ignored. This can shadow other errors which we may not have expected and could lead to possible bugs

Another big issue is when error occurs in loops or If Else statements. This could create unexpected flow of code and also infinite loops

Andrew: Can you show an example code to demonstrate the last issue?

Me: Yes, if we see this code

And I thought I knew QTP!

```
On Error Resume Next
   x = 0
   If x = 1/x Then
       Reporter.ReportEvent micPass, "The test has passed", "Passed"
   Else
       Reporter.ReportEvent micFail, "The test has failed", "Failed"
   End If
On Error Goto 0
```

If run the above script the test would pass even though our condition has failed because of Division by zero. When the error occurs we move into the next line of code which is to pass the test case. This approach results in erroneous results

Andrew: How do you fix this then?

Me: One possible way is to test the negative conditions first instead of the positive condition. So we can change the code as below

```
On Error Resume Next
   x = 0
   If x <> 1/x Then
       Reporter.ReportEvent micFail, "The test has failed", "Failed"
   Else
       Reporter.ReportEvent micPass, "The test has passed", "Passed"
   End If
On Error Goto 0
```

In this case even if the error occurs we will move into the micFail statement which would make the script fail and call further analysis.

Another fix would be as I had pointed out earlier that we can move the code inside a Function and then call the Function.

94   Tarun

Personal Interview–Round 1

Andrew: Can you show an example of a situation where you expect the error and check for it?

Me: Sure. Consider that we open a text file and that text file doesn't exist. In this case we will get an error. Now we know that our code can probably raise such an issue so we need to handle this.

```
Set FSO = CreateObject("Scripting.FileSystemObject")

Set oFile = FSO.OpenTextFile("C:\FileNotExist.txt")
sContent = FSO.ReadAll
oFile.Close

Set oFile = Nothing
Set FSO = Nothing
```

When we run the above code if the file doesn't exist, an error will be thrown. Now, to handle this situation we can debug the code and note down the error number. We can then update the code as below

```
Set FSO = CreateObject("Scripting.FileSystemObject")

'Enable error handling
On Error Resume Next

'Clear any existing error
Err.clear

'Open the text file
Set oFile = FSO.OpenTextFile("C:\FileNotExist.txt")
```

```
'File not found error
If Err.Number  = 53 Then
   Reporter.ReportEvent micFail, "File not Found", "Failed to find the file - C:\FileNotExist.txt"
   ExitRun
End If
'Disable error handling
On Error Goto 0
sContent = FSO.ReadAll
oFile.Close
```

Andrew: What is a recovery scenario?

Me: Recovery scenario is a QTP feature which allows a test to recover from unexpected errors. There are 3 parts of a recovery scenario – a trigger, a recovery Action and a post recovery Action. A trigger can be of 4 types

- *Object state*
- *Popup window*
- *Application crash*
- *Test Run Error*

Then a recovery Action could be of 3 types

- *Mouse/Keyboard Action*
- *Function call*

- Close the application process
- Restart windows

Post recovery has different options

- Restart the test
- Move to next step
- Move to next Action iteration
- Move to next test iteration
- Stop the test
- Re-run the step

Andrew: My script shows an error Dialog with Debug, Retry, Stop and Skip button. Basically the QTP's error dialog window. Can we handle this using Recovery Scenario?

Me: Once the error is raised recovery scenarios cannot be used to handle error dialog as QTP waits for user inputs. Recovery scenarios can be configured to run in test with below options:

- On every step
- When error occurs
- Never

QTP will only raise the error dialog if the setting has been set to Never Execute Recovery Scenario or none of the recovery scenario triggers are satisfied.

Andrew: How are recovery scenarios different from 'On Error Resume Next' then?

And I thought I knew QTP!

Me: 'On Error Resume Next' is helpful where we know the location and the type of error. E.g. If we are opening a file then we can expect 2 possible errors: one is file doesn't exist and other is an access denied error. We can use On Error Resume Next for these checks with 'Err.Number' and use specific numbers associated with these errors to take the next Action.

Recovery scenarios are used for unexpected errors where the location of the error is also unexpected. E.g. While testing a web application, we get a security information popup between one of the navigations. In this case, we can use a recovery scenario to take care of it.

Andrew: When I navigate to my web application, I get a security popup error. Now this only happens when I am navigating to the URL, so the location of the error and type of the error is very well know here. Once I login to the application if the user id or password was wrong I get an error dialog for the same. Now would you handle these dialogs in script code as the location and type is known or would you use Recovery Scenario?

Me: Well, we can handle it both the ways but if you ask my preference then I would handle the certificate error using Recovery Scenario and the invalid password dialog using conditional statements in my code.

Andrew: Any specific reason for that approach?

Me: I prefer the scripts to be concerned about the application rather than such environment issues. Since a password window is application related, I would want my scripts to handle it.

Andrew: I want to fire the Recovery Scenario but not on error, what can I do?

Me: We use the option to execute the recovery scenario on every step or we can use 'Recovery.Activate' to fire the recovery scenario in code.

Personal Interview–Round 1

Andrew: Consider the below code

```
On Error Resume Next
Window("hwnd:=0").Click
Print "Recovery scenario not fired"
On Error Goto 0
```

Now, I have a Recovery Scenario associated with the above code with trigger set as any error and the recovery Action as Function a call which prints 'Recovery Scenario fired' and my post recovery Action is to stop the test. What would happen in this case? Which message will I get?

Me: (ummm....that's a tough one...On Error Resume Next should not let any error to be raised, no error means no Recovery Scenario also. But if that was the case just using On Error Resume Next will disable the Recovery Scenario feature all together)

I have not tried this before, so I am not sure I can give you a correct answer but I can try and answer in a logical way. Logic says that if just using On Error Resume Next disables the recovery scenario feature then this could be a huge gap in design, so I feel that we should get 'Recovery Scenario fired' message.

(...nice question...this is becoming fun now...)

 Note: The recovery scenario will be fired even if 'On Error Resume Next' statement is used in the code

Andrew: What are the limitations of Recovery scenarios?

Me: Few limitations that I am aware of are

- They run in the same thread as the script, so if a Modal dialog blocks a script then recovery scenario can't be fired

- Recovery scenarios don't work on VBScript errors which don't involve QTP's Test Object. So for 'x = 2/0' statement, a recovery scenario will not be fired.

- Another problem with recovery scenario is the chaining of Recovery scenarios. If I have 2 scenarios whose trigger is satisfied then the first in the list is executed and then the next one. There is no way for me to stop that chain on a specific scenario

- We cannot add recovery scenarios to a test at run-time

- There is no support for Automation object model APIs to be able to see what recovery scenario a file has.

Andrew: You just now spoke about Automation Object Model. What is it?

Me: QTP Automation Object Model, which is an acronym for AOM is a COM based library which exposes QTP's UI related Functions through different interfaces.

We can use QTP AOM to automate tasks like opening a Test, running it, setting the result location, configuring script options etc from outside of QTP through VBScript or compiled code.

Andrew: I want to execute a DOS command, consider IPConfig as the command. How would I execute the same from QTP and capture the results?

Me: There are different ways of doing this. One way is to use the Exec method of WScript shell which provides access to the StdOut object. This was we can read what was outputted from the command, but this approach doesn't work well if we have an interactive command like FTP or something.

Another way to execute the command is to move all the output to a file using the redirection operator. So by executing 'ipconfig >C:\ip.txt' ms-dos will move all the output to the ip.txt file.

Andrew: How would you execute this in QTP?

Me: I would use SystemUtil for this. The command would be

```
SystemUtil.Run "cmd", "/C ipconfig >C:\ip.txt"
```

Or we can also use the run method of WScript Shell

```
CreateObject("WScript.Shell").Run("cmd /C ipconfig >c:\ip.txt")
```

Andrew: Ok, any other ways of executing DOS command you wanted to tell me?

Me: Yes, another option is to launch the MS-DOS command window and then use the 'Window.Type' method to send commands to the window opened. In this case we have few limitations like we will have to use GetVisibleText method to read the command output. The GetVisibleText method can be unreliable on certain machines and it will only give the text available on screen. If a command outputs text more than the screen size then the GetVisibleText will capture only the visible part. This would mean we will get on partial text of the output

Andrew: How would you execute a UNIX command?

Me: QTP doesn't directly support execution of UNIX commands. So we need to use some 3rd party tool to connect to the UNIX box and then automate the UI of that 3rd party tool through QTP. Putty is usually the most common tool used in such cases. We can also use COM based libraries which allow connecting to UNIX box and then executing commands.

Andrew: Can you name any such COM Based libraries?

Me: I have not used one till now as we have used Putty only

Andrew: What is a DataTable?

Me: A DataTable is a similar to an Excel file that allows driving data to a test case for a several number of rows. Each Action has its own copy of a tab known as Local DataTable. The whole test shares a common Global DataTable.

Andrew: How do we configure Iterations of DataTable?

Me: Iterations can be done on two entities: Test and Action call. For setting Test iterations, we can go to Tests->Settings menu and then set the iterations in the run tab. The options are to be run on first row, run on all rows and run from row X to Y.

For setting Action iterations, we can right-click on the Action call in the Keyword View and then set Action Call Properties to have iterations.

Andrew: How do you find some data in DataTable?

Me: There is no method in DataTable object that lets us find data. The only way is to enumerate each row one by one and then checking for the desired value. DataTable provides a ValueByRow method to get value from a specific row.

Andrew: How do you delete a row from DataTable?

Me: We can right click the row and the select delete option. But there is no run-time method in QTP for doing that. We can blank out the values in a Row through code. A workaround for this would be to export the sheet and then open the same through Excel COM API, delete the row and the re-import it. But this would be a very poor approach.

Andrew: How do you export the DataTable on specific intervals in Test?

Me: By intervals, do you mean time intervals or iterations?

Andrew: Assuming either of the two?

Me: We can create a Function that we keep on calling. It would have some counter that we will keep on incrementing then once it reaches the interval we would export the Data Table

```
Dim iCounter: iCounter = 0

Function ExportTableOnInterval(ByVal Interval)
```

Personal Interview–Round 1

```
    iCounter = iCounter + 1
    If iCounter > Interval Then
        iCounter = 0
        DataTable.Export "C:\DataTableDump.xls"
    End if
End Function
```

Andrew: In what sort of situation would you need to export the Data Table?

Me: One situation that I can think of is that when the Test runs for long hours and the results of individual rows are stored in DataTable. In such situation, a crash of system or QTP can lead to loss of results, so to minimize the loss we can export the Data Table on regular intervals.

Andrew: Can we save a QTP script as a single file?

Me: QTP has an option to export test to zip file. But it can't be opened without extracting its files to a folder using the import option.

Andrew: How do you schedule a Test run at a specific time?

Me: We can use the scheduling feature of QC. In TestLab's test flow view we can add a condition on a test to be executed on the specified time. But for this we need to initiate the run of the test and keep QC running, QC won't initiate the run on by itself.

Another option is to create a VBS file which uses QTP AOM to run the test and then launch that using windows task scheduler.

 Note: We can create a VBS file that can open the test and execute the same

```
Dim qtApp 'As QuickTest.Application ' Declare the Applica-
tion object variable
```

```
Dim qtTest 'As QuickTest.Test ' Declare a Test Object variable
Dim qtResultsOpt 'As QuickTest.RunResultsOptions ' Declare a Run Results Options object variable


Set qtApp = CreateObject("QuickTest.Application") ' Create the Application object
qtApp.Launch ' Start QuickTest
qtApp.Visible = True ' Make the QuickTest application visible


' Set QuickTest run options
qtApp.Options.Run.ImageCaptureForTestResults = "OnError"


qtApp.Options.Run.RunMode = "Fast"
qtApp.Options.Run.ViewResults = False


qtApp.Open "C:\Tests\Test1", True ' Open the test in read-only mode


' set run settings for the test
Set qtTest = qtApp.Test
qtTest.Settings.Run.OnError = "NextStep" ' Instruct QuickTest to perform next step when error occurs
```

```
Set qtResultsOpt = CreateObject("QuickTest.RunResultsOp-
tions") ' Create the Run Results Options object

qtResultsOpt.ResultsLocation = "C:\Tests\Test1\Res1" ' Set
the results location


qtTest.Run qtResultsOpt ' Run the test


MsgBox qtTest.LastRunResults.Status ' Check the results of
the test run

qtTest.Close ' Close the test


Set qtResultsOpt = Nothing ' Release the Run Results Options
object

Set qtTest = Nothing ' Release the Test Object

Set qtApp = Nothing ' Release the Application object
```

We can configure the above script to read the script names or paths from an external file and make the driver script flexible based on our needs. Once the driver script is ready we need to create a task in the task scheduler. To create a new task open control panel-> Scheduled Tasks->Add Scheduled Task.

In the wizard click on the Browse... button and select the wscript file from the <Windows>\system32\ folder. Once the task is created right click on the task and open the properties.

And I thought I knew QTP!

Add the path of the file in the Run text box as shown in the image below.

The task is now ready and can be tested by right clicking and choosing the Run option.

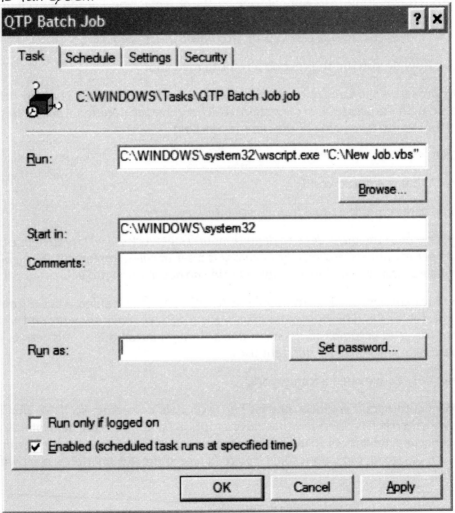

Andrew: How do you run multiple tests in one go?

Me: We can run multiple tests but only one by one on a machine. For this we can use QTP's AOM to create a script which loops through each test, opens the test, runs it, captures the result and moves to the next test in order to repeat the procedure.

Andrew: How do you run the test on a remote machine?

Me: The process is the same as of running the test on a local machine. The only thing that changes is with the CreateObject Function, which receives an additional parameter, which is the IP or the machine name of the remote machine.

```
Set QTP = CreateObject("QuickTest.Application", "RemoteIP")
```

Andrew: What is a framework?

Me: (huh! here comes the usual framework interview attack)

Well there are not too many defined definitions of framework so I will try to define the best I can. As per my understanding a Framework is a set of rules, implementation guidelines and standards that you define for implementing an automation project.

Also, in my experience, different people have different ideas behind what a framework should be. I think this is because most of us work in different environments and with users that are trying to accomplish dissimilar goals through automation. This results in each user creating frameworks for different uses.

Andrew: Why do we need a framework?

Me: Every automation engineer has his/her own style of coding, so if we don't use a framework then in few days we would have scripts in hand, which are developed in an ad-hoc manner and are difficult to maintain over a longer period of time. Lack of a framework or a poorly designed framework could lead to failure of the Automation project with little to no ROI.

Andrew: What are different types of framework?

Me: There are three key types of frameworks namely Data Driven, Keyword driven and Hybrid framework.

Andrew: What about Modular frameworks, functional decomposition, Action based frameworks? Haven't you heard about them?

Me: Yes, I have used them also. But, these are generally subsets of the above three.

Andrew: Tell me something more of the three key frameworks you listed.

Me: Ok, I will start with the Data Driven then. In Data Driven frameworks, relevant application data plays a key role in execution and as the name suggests, data drives the whole execution. This is useful when the flow of the test remains the same but data keeps on changing. A very simple example would be a calculator application where we can have a value1, operator1, value2, operator2 and result. We can have multiple rows of data and the test can execute all the rows and report the results for each row

In Keyword Driven frameworks, keywords play a major role and decide the execution flow. There are two main types in these frameworks as well – generic keyword driven and application specific keyword driven. Generic frameworks are created to be adapted to different sorts of applications and technologies. Application specific KD frameworks are mainly intended to automate a specific type of application and are generally not transferrable to other technologies or applications.

Also, Generic keyword driven frameworks are ones where we create keyword for normal operations and they are generally close to English terms. For example:

```
Open "IE"
Open "www.google.com"
Enter "q", "Test This"
Click "Google search"
```

As you can see the actions are written using generic keywords and are more like English actions. The advantage of such frameworks is that the script developed in such frameworks can be ported to a different tool by implementing the core framework for that tool. Even though it seems advantageous to take such an approach, I personally feel this is very impractical as it takes away the advantage which the Automation tool might be offering. Also, the framework development and unit testing becomes a huge project of its own. And porting of scripts from one Automation tool to another automation tool can be taken as a onetime activity.

The other type of Keyword Driven frameworks are the application specific keyword driven framework. In these frameworks the keywords are related to application. E.g.

```
BANK_Launch
BANK_Login "user", "Password"
BANK_MakeTransaction "FromAccount", "ToAccount", "Amount"
BANK_VerifyLastTransaction "FromAccount"
BANK_Logout
```

In the above code, the keywords are specific to application and perform certain actions on the application. The advantage of this framework is that once all the keywords and their input and output parameters have been fixed, the details can be shared with the application SME's. These SME's then with some basic knowledge of the automation tool can build upon the scripts. These keywords provide abstraction to the users as they don't need to know the internal working or complexity of the Functions.

The Functional Decomposition and Action based frameworks that you mentioned are used to implement an application specific framework where the keywords are implemented using Functions or Actions. Therefore, they are more of a subset of Keyword Driven framework.

Andrew: What does a keyword mean in a Keyword Driven framework?

Me: A keyword is just a short form or a name associated with some action or operation or an object.

Andrew: What about Hybrid frameworks?

Me: When we combine the Data Driven and a Keyword Driven framework then framework becomes a hybrid framework. These are frameworks which are commonly used these days and most frameworks finally evolve to a hybrid framework. For example, consider the above Keyword-Driven framework approach that uses keywords to drive application specific actions. Now, this framework can be linked to a data-source keeping the same keywords to become a Keyword Driven and Data-Driven framework – a hybrid.

Andrew: I was wondering, you said Modular, Functional decomposition or Library driven and Action based frameworks are subsets of Data-Driven, Keyword-Driven and Hybrid frameworks. Why so?

Me: I say that because regardless of the type of framework you create, it eventually runs on one of the concepts from either a Data-Driven, Keyword-Driven or a Hybrid framework. For example, let's consider an Action based framework. Every Action represents a keyword and when we call these Actions in some sequence it becomes a Keyword Driven approach only.

Andrew: Agreed.

Andrew: What about the functional decomposition or a library driven structure?

Me: The same concept applies to them. If this type of a framework uses data from an outside data-source, then in the end, it would be called a Data-Driven framework.

However, if this framework contains Functions that simplify the way scripts are created, for example, the login Function has a Login keyword that accepts the username and password strings; or another Function that enables adding items to the cart, or a Function that enables logging out of the application:

```
Login MyUser, MyPassword
AddItemToCart Array("Book", "Candy")
Logout
```

This would then be considered more of a Keyword-Driven approach than a Library driven approach. Also, if the data comes from a data-source here, then it would become a Hybrid structure instead of a simple functional decomposition structure.

Andrew: Why is QTP call Advanced Keyword Driven Testing tool?

Me: That name comes from the Keyword View where each operation is represented by a keyword object and a keyword operation on them. Though in my personal opinion, I won't call it a Keyword Driven tool. Changing QTP to a pure keyword driven tool will bring about various limitations and would also result in much redesign of the overall implementation.

Andrew: Have you heard of the Object Repository framework?

Me: Well, I have heard about many weird names that just don't exist but these names are generated somehow. This is usually because there is not much documentation available for the same. I say this because an Object Repository does not provide the guidelines to automate an application. It's simply a storage area for QTP's Test Objects.

Andrew: Well we spoke about different types of framework but what actually makes a framework?

Me: Well framework is a lot more complex than their definitions. The reason being our current definition about the frameworks, one only talks about the execution model of the same and it doesn't talk about other key parts. There are few keys factors a framework needs to consider

- Reporting model
- Execution model
- Batch execution model
- Test Data management and maintenance

Personal Interview–Round 1

- *Script maintenance process*
- *Object repository maintenance*
- *Naming/Coding guidelines or standards*
- *Support for multiple version of application*
- *Estimation guidelines/process*
- *Error or Exception handling*
- *Reusability*
- *Version control*
- *Ease of use*
- *Documentation*
- *Logging and Debugging*

Andrew: Could you please elaborate what have you done for 'Logging and Debugging' part in your frameworks?

Me: For logging we have developed a specific class that does the logging. So each method of ours looks something like below

```
Function MyTest(ParamA, ParamB)
   Dim OC
   Set OC = Log.EnterFunction("MyTest", _
        Array("ParamA", ParamA, "ParamB", ParamB))
   'Function code
End Function
```

And I thought I knew QTP!

The EnterFunction method of the Log class writes all the details about the Function name and its parameters into a debug text file. We pass the names of the parameter as one array and values of those parameters as another array. The Function returns an object. The object is stored in a local variable OC. This approach is helpful in capturing the exit from the Function because when the Function is exited whether intentionally or through an error the Class_Terminate gets fired on the OC object class. This captures the exit from the Function and is used to decrease the indentation depth in the log file.

Note: For readers convenience the code pertaining to above example is shown below

```
Function Repeat(sText, iCount)
   Dim i
   For i = 1 To iCount
      Repeat = Repeat + sText
   Next
End Function


'Class to define a function call
Class FunctionCall
   'The name of the function
   Dim FunctionName


   'Array of parameters. Name and value pair
   Dim Parameters
```

```vbscript
'Time of the call
Dim CallTime


Sub Class_Initialize()
   CallTime = Now()
End Sub



Function GetCallDetails()
   Dim s_Call, s_Params

   'Check if parameter is an object
   If IsObject(Parameters) Then
      s_Params = "[object:" & TypeName(Parameters) & "]"
   ElseIf Not IsArray(Parameters) Then
      'If not an array convert it to a string
      s_Params = CStr(Parameters)
   Else
      'We assume the parameters are key value pairs
      'Make sure we have an even number of elements in array
      If (UBound(Parameters) - LBound(Parameters) + 1) Mod 2 = 0 Then
```

```
            Dim j
            s_Params = ""
              For j = LBound(Parameters) To UBound(Parameters) Step 2
                s_Params = s_Params & Parameters(j) & ":="
                'Check if the value of the parameter is a object
                If IsObject(Parameters(j + 1)) Then
            s_Params = s_Params & "[object:" & TypeName(Parameters(j + 1)) & "] ,"
                Else
                  s_Params = s_Params & GetArrayText(Parameters(j + 1)) & " ,"
                End If
              Next
            Else
              s_Params = "[Error key value pair not specified]"
            End If


            If Right(s_Params, 1) = "," Then
              s_Params = Left(s_Params, Len(s_Params) -1)
            End If
          End If
          s_Call = FunctionName & " (" & s_Params & ")"
```

```
    GetCallDetails = s_Call
End Function



Private Function GetArrayText(ByVal Arr)
On Error Resume Next
Err.clear

If IsArray(Arr) Then
  Dim newArr
  newArr = Arr

  Dim i
  For i = LBound(newArr) to UBound(newArr)
  if IsObject(newArr(i)) Then
    newArr(i) = ""
  Else
    newArr(i) = CStr(newArr(i))
  End if
  Next


  GetArrayText = "Array(""" & Join(newArr, """,""") & ")"
```

```
      Else
         GetArrayText = Arr
      End if

      If Err.number Then
         GetArrayText = ""
      End If
   End Function
End Class


'Function to get new instance of the function call
Function NewFunctionCall()
   Set NewFunctionCall = New FunctionCall
End Function


'Class to get a callback executed. We need to set the two
'members
'Caller - The object which needs the callback
'CallbackCode - Code to be executed for callback
Class Callback
   Public Caller
   Public CallBackCode
```

```
    Sub Class_Terminate()
        Execute CallBackCode
    End Sub
End Class


'Function get a new call object
Function NewCallback()
    Set NewCallback = New Callback
End Function


Dim DEBUG_LOG
DEBUG_LOG = True


'Class logger allows logging function calls and entering log text
'in between

Class Logger
    'Dictionary to maintain current stack trace
    Private oStackTrace
    Private sLog
```

'Class initialization

```vbscript
Sub Class_Initialize()
    Set oStackTrace = CreateObject("Scripting.Dictionary")
    sLog = ""
End Sub


Function SaveDebugLog()
If DEBUG_LOG and sLog<> "" Then
    Dim FSO, sFile, debugFile
    Set FSO = CreateObject("Scripting.FileSystemObject")
    sFile = "Debug_" & Replace(Replace(Replace(Now(), ":", "_"), "/", "_"), " ", "_") & ".txt"

    Set debugFile = FSO.CreateTextFile(Reporter.ReportPath & "\Report\" & sFile, True)
    debugFile.Write sLog
    debugFile.Close
    Set debugFile = Nothing
    Set FSO = Nothing
    sLog = ""
End If
```

```
End Function


'Class termination
Sub Class_Terminate()
SaveDebugLog
   Set oStackTrace = Nothing
End Sub


'Private functions to Push and Pop function calls
Private Function Push(oFunctionCall)
   sLog = sLog + "[" & oFunctionCall.CallTime & "] " & _
       Repeat(" | -", (oStackTrace.Count) * 2) & _
     " Start Function - " & oFunctionCall.GetCallDetails _
         & vbNewLine
   Set oStackTrace(oStackTrace.Count + 1) = oFunctionCall
End Function



Sub Write(ByVal sText)
   sLog = sLog & "[" & Now() & "] " _
       & Repeat(" | -", (oStackTrace.Count)) _
       & vbTab & sText & vbNewLine
```

```
    End Sub


    'Private function to pop and log the end of last function call

    Private Sub Pop()
        Dim oLastCall
        'Get the details about last function call
        Set oLastCall = oStackTrace(oStackTrace.Count)

        'Remove the last function from the stack
        oStackTrace.Remove oStackTrace.Count

    'Append the end of function to log
        sLog = sLog + "[" & oLastCall.CallTime & "] " _
            & Repeat(" | -", (oStackTrace.Count) * 2) _
            & " End Function - " & oLastCall.GetCallDetails _
            & vbNewLine


        Set oLastCall = Nothing
    End Sub
```

```
'Function to pop the last function call

Sub LeaveFunction()
   Call Pop
End Sub


'Method to be called when entering the function
'FunctionName - Name of the function being called
'Parameters - Array of key value pair
Function EnterFunction(FunctionName, Parameters)
    'Create a new function call with given function name
    'and paramates
    Dim oFuncCall
    Set oFuncCall = NewFunctionCall
    oFuncCall.FunctionName = FunctionName
    oFuncCall.Parameters = Parameters

    'Push the function call on to the stack
    Push oFuncCall

    'Create a new callback
```

```
    Set EnterFunction = New CallBack


    'Set the caller as current object
    Set EnterFunction.Caller = Me


    'Set the callbackcode to execute leave function
    EnterFunction.CallBackCode = "Caller.LeaveFunction"
End Function

Function Report()
Set Report = New CallBack
Set Report.Caller = Me
Report.CallBackCode = "Caller.SaveDebugLog"
End Function

Function GetLog()
   GetLog = sLog
End Function

Function PrintLog()
   Print "- - - - - - - - - - - - - - START LOG - - - - - - - - - - - - - - - - -"
   Print GetLog()
   Print "- - - - - - - - - - - - - - END LOG - - - - - - - - - - - - - - - -"
```

```
   End Function

   'Function to get the current stack trace4

   Function GetStackTrace()

      Dim i

      Dim s_TraceLog, s_CurrentFunction

      s_TraceLog = ""

      For i = 1 To oStackTrace.Count

        s_TraceLog = s_TraceLog & "[" & oStackTrace(i).CallTime
& "] -" & String((i -1) * 2, "-")

          s_TraceLog = s_TraceLog & oStackTrace(i).GetCallDe-
tails() & vbNewLine

      Next

      GetStackTrace = s_TraceLog

   End Function

   'Function to print the stack trace

   Sub PrintStackTrace()

      Print "- START STACK TRACE -"

      Print GetStackTrace()

      Print "- END STACK TRACE -"

   End Sub

End Class
```

And I thought I knew QTP!

```
Dim Log
Set Log = New Logger
```

Andrew: What about version control?

Me: We have used QC in all projects and we didn't have version control enabled on the QC server. So we used to capture all our libraries, environment, data files in VSS. At the end of day we checkout all files from VSS to local and then run a script to update the files on local with the ones in QC. Once the local copy is in sync with QC we used to check in everything back to VSS.

Andrew: What about naming and coding guidelines? What kind of guidelines you had in this case.

Me: We had defined how variables need to be declared. We had prefix define for each type of data type. For example variables with int will start with 'i', objects with 'o', long with 'l', string with 's' etc...

Similarly for environment variables we had a convention that the name would be in all Caps and prefixed with 'ENV_'. For Function created in the library we had an convention any generic Function would start with 'GEN_', framework related Function would start with 'FRM_', application related Functions would start with 'APP_' and so on. There were many such conventions that we had defined and documented for our framework

Andrew: What about Batch execution model? How did you do that?

Me: Again in most of the projects we had we used QC TestSets to drive the execution. We would just filter out the tests based on our favourites and run those test in a batch. There were no driver scripts in most of the cases.

But for one of the projects we had to run scripts from local only, so for that we had developed an Excel macro where we can add the script names and their location in the Excel, after selecting multiple rows we used to click the Run button macro which in turn would figure out the scripts to be run and will run those scripts. We also had an additional feature of specifying a specific machine or setting it to 'Any' for the macro to choose one of the lab machines. It would then after every 1 minute query the status back from all running scripts. The status would then be reported back to script and any failures at the end were also captured in an Excel file.

Andrew: What is a Driver script?

Me: A Driver script is a script which drives the whole execution of a test suite. Based on how the framework is designed, the Driver script may have different tasks to be performed. The simplest of the Driver scripts will just open the test and execute it. A complex Driver script might have tasks like adding Recovery Scenarios, Object Repositories, Library Files, cleaning up databases, setting up test data, updating results to database, QC or Excel etc.

In other words, a Driver script can also be defined as an engine that controls the way tests are executed, taking into consideration all components and layers of the framework involved in the overall implementation.

Andrew: What are Stubs?

Me: A Stub replaces an actual functionality by dummy calls or dummy actions. Consider a case where we have an upgrade promotion page which is shown to user when they order a product. Now, our test cases will contain some code to handle this page. For some reason developers take this page out from the flow for fixing an issue, in that case instead of removing the code for handling the upgrade page, we can stub it out where the stub only does that part which is necessary for the script to work and doesn't take any actual action on the application.

Andrew: Well you have written that your current project was on BPT. I have asked you about frameworks and you haven't even mentioned BPT as one. Can you please tell me a little bit about your work with BPT as well?

Me: No, I was just discussing frameworks in general and that's why I didn't mention BPT. It is not a type of framework that users like us create. It is a proprietary implementation from HP which works tightly with QTP and QC integration.

Andrew: Hmmm, What is BPT?

Me: BPT is Business Process Testing framework from HP. It is built on the concept of separating automation work between Application SME's and Automation engineers. BPT at core has re-usable components which can be pulled into a business process to create a automated Test. Application SME's can define what these components would be and what would be their functionality while the Automation engineers implement how the functionality works on to the application using QTP.

Andrew: What are the advantages of BPT?

Me: As I just mentioned earlier that BPT separates work for Application SME's and Automation engineers. Once the basic components are ready even manual testers can create automated test cases. Also the BPT UI is a bit flexible to implement common changes to components. E.g. Adding additional parameters to components doesn't need any modification to the Business Process as the changes are propagated automatically.

Andrew: How is a Business Process Test different from a normal QTP Test?

Me: It is very different from a QTP Test. Each component in a Business Process serves or rather runs as an individual test. So the execution happens through running of series of components which behave the same as a individual QTP test. This approach of BPT also hampers performance at times when the QC connection is not very fast as different components gets download to our local machine and then a component is stopped completely to load the new component.

Andrew: What are the disadvantages of BPT?

Me: BPT framework puts few restrictions on the way tests can be designed. This can appear to be a disadvantage to some users but a good framework design over BPT can still fix a lot of things up. But few common issues that I have faced with BPT are:

- *Performance can become poor if too many components are used is a Business Process*

Personal Interview–Round 1

- There are no Business Process/Test level parameters which makes it tedious to support global Test level variables. We can still use a few workarounds

- Error handling of BPT is poor. If an error occurs in BPT component then QTP will continue to the next line and I was surprised to see there is no option to change that behaviour

- On Error option of BPT only allows us to stop The test or continue, it doesn't allow us to end the Test and execute a cleanup component

- Then there are few UI related things which if would have been implemented in a better way would have helped in increasing efficiency

Andrew: What kind of a framework is BPT?

Me: BPT is a hybrid framework. It is primarily used as a Keyword Driven framework. Where every component used in the Business Process serves as a keyword. We can data drive these components through component or group iterations or we can also iterate the whole test through iterations in test lab.

Andrew: How in your project did you manage data inputs to your BPT scripts?

Me: Well what we used to do was that in components whichever input parameters were not dependent on test-data, we used to hard-code them. This way whenever we called the component in test we had the default values for such parameters. In the Business Process if the value for the parameter was specific to the test we used to hardcode the value, if not then we used to map the same to the run-time parameters. This way the user only had to enter the run-time parameters which are necessary to run the test.

Andrew: I want to data drive a Business Process test case through external data file, how would I do that?

Me: The only way I know of is to go to Test Lab and the import the iterations manually from a CSV file.

Andrew: How would you do it through code?

Me: There is no way that I am aware of for doing this using code. As far as my knowledge goes QC/BPT doesn't provide any API for this

Andrew: Suppose we have 10 libraries associated in my Application area with their QC Paths. Now we want an option to use these libraries from a local path. What can be done?

Me: Unlike normal Tests in BPT's Application Area we can't use relative paths. I am not sure of the idea behind HP not providing support for that but we can't use relative paths. The one reason I can think of could be the support for Keyword only components which can be created in QC only also. The only possible workaround I could see would be to create another application area which has all paths associated as local paths. When we want to change to local files we can temporarily change the application area of the current component and change it back to original once our work is done.

Andrew: Can we disable Keyword View in QTP?

Me: This is possible only in QTP 11.

Andrew: How do you measure time taken by a certain operation?

Me: There are few possible ways to do it. One would be to store the value returned by the Timer function and then subtract the old value from new Timer return value to see the time taken. But this doesn't work well if the clock goes past midnight. So a better approach would be to capture the date and time itself using the 'Now' method and then use DateDiff after the operation to the find the difference between current time and time capture before the operation. There also exists the MercuryTimers utility object which we can use to start and stop a timer.

Andrew: How do you use this MercuryTimers object?

Me: There is a start and stop method available for a timer object. We can use it in this fashion

Personal Interview–Round 1

```
MercuryTimers("Test").Start
Wait 5
MercuryTimers("Test").Stop
Msgbox MercuryTimers("Test")
```

Andrew: Can we use this approach to measure load time of a website and do a bit of performance testing?

Me: *We can certainly use this approach but the data generated from these tests won't be accurate as load time will also include the process related times.*

Andrew: So how do you measure such time?

Me: *We should use a tool which is specifically built for this. Something like LoadRunner would be appropriate for such task.*

Andrew: Nurat, We are done with this round of interview. You can wait outside now.

Finally!! Round one of the PI was over. I was feeling very hungry and wanted to run for my lunch. But I didn't know if I would be called for another round of interview or would be rejected in this one. I doubted my rejection though as the interview had gone fine and I managed to answer most of the questions.

I kept on waiting for the next action and there was no response from anyone for another fifty minutes. I was starting to feel frustrated, not because of the waiting but mostly because of the intense hunger. After ten minutes Andrew came and told me that there will be a next round and it will start in the next thirty minutes.

I left the room and started looking for a cafeteria. I decided to ask the lady at the reception. I went back to her and asked for the way to the cafeteria.

'Hi, can you please tell me the way to cafeteria?' I asked.

'Oh sure... Go straight then take left from there take staircases and you will find yourself in cafeteria.' She showed the way. I thanked her in response.

I rushed to the cafeteria and got a non-veg meal for myself. Though in my family we were not allowed to eat non-veg on Monday, Tuesday and Thursdays but after joining work I decided to let go of all these restrictions.

It is difficult to get good food when you are away from the family and not having non-veg just cuts down many good options. The meal was nice but I couldn't enjoy the taste of it as I was worried about my next round. I quickly finished my meal and got ready for my next round.

Personal Interview–Round 1

I reached back on time and waited outside the meeting room. But as usual, things always seemed to be delayed here. After yet another forty-minute delay I was asked to wait in the meeting room. Luckily this time I didn't have to wait inside the room.

The interviewer entered the room. He looked very experienced and professional.

Face to Face Interview

Round 2

⚘

Alex: Hi Nurat, I am Alex. I am the Sr. Technical Architect here and I look after the framework and few other teams. This would be more of a practical interview. You do have a laptop in front of you with QTP 10 installed. You can refer to the same whenever I say so.

Let's start.

I understand that limitations of any product are generally subjective. What one sees as a limitation may not be seen in the same light by another user. So in your opinion then, what do you think are the limitations of QTP?

Me: Well, I think it is difficult to list what QTP can't do instead of what it can. Let me see if I can put few limitations or rather things that QTP can't do:

- QTP cannot automate applications running on a remote machine.

- It doesn't allow running a script without QTP installed on the machine.

- It has very limited support for VC++ or C++ applications.

- The documentation of QTP doesn't show practical uses of its features at times. They just show how to use the feature. We can't always expect companies to put a lot of effort in this direction due to strict deadlines and the complexity of task at hand. It is the user community that holds a greater responsibility here. There have been recent efforts with several user groups as well the latest book from Tarun Lalwani

- 'QuickTest Professional Unplugged'. So, this gap is certainly being minimized with efforts from users around the globe.

- QTP supports only 3 Web browsers: IE, Firefox and Netscape. Support for other browsers like Chrome, Safari for Windows, and Opera etc. is still not available. This hinders clients' efforts of utilizing automated suites across multiple browsers.

- QTP uses VBScript as its core language which itself has lots of limitations. Support for .NET languages can help make QTP far more powerful then it currently is.

- The way QTP is marketed may also be seen as a limitation. It is marketed as a tool that can be used by individuals with no development knowledge. This creates an impression in the user's mind that knowledge of development isn't necessary. But in my opinion, the less knowledge we have of development technologies, the more limitations we will believe QTP to have. This is mainly because our vision is restricted to a smaller box and we can't see so many easy solutions outside the box.

- Also, there are features like Object Repository management and maintenance, error handling capabilities etc which can be greatly improved.

There would many others that I can think of but I guess my answer won't end. I hope these points do answer what you were looking for.

Alex: Yes it does indeed. Okay, tell me what challenges have you faced in QTP Test Automation and what have you done to overcome them?

Me: (This long interview seems to be the biggest one till now.... ☹, anyhow...)

Every project that I have worked in has brought different kinds of challenges.

The first project that I worked on had two key challenges: one was that objects in the application used to keep on changing and second was with the data that we used lacked the required plans available for ordering. For the dynamic object handling issue, we created a new Excel based Object Repository which allowed adding multiple definitions of an object which are then resolved at run-time. This enabled us to support similar objects on different pages with different properties under a single name. This made scripting a lot easier and maintenance less painful.

For the test data issue, we created data validation scripts which go into the flow and capture various issues with the data and available plans. This way, we were able to bring down failures due to bad test data by over 75 percent.

In another project of mine, our team had to perform automation of an asynchronous ordering system. Within the application, once the order is placed, the time for it to be processed fluctuated. It was completely variable and out of our control. We couldn't make the script wait for a high timeout as it would take away a lot of precious time from our execution window. So, we created one script which places the order and another (polling) script which ran twice a day to check for order status and updated its execution status in QC.

Another challenge I faced was on a project where we had to download data by using different user accounts and merge it into one Excel Worksheet. There were two issues we had in automating this scenario. First was performance; because we were using GetCellData and fetching data cell by cell into Excel it was taking ages to be completely downloaded. We found out that if we simply copy the cells by sending CTRL+A, then CTRL+C, we can paste the whole data directly in Excel. This reduced the time to a mere 2 sec from over 20 minutes. Second issue was with Explorer constantly crashing. We found out that, because we used to close Explorer through our script, it would crash IE and cause the script to hang. This also prevented our recovery scenarios to work correctly. So, we decided to logout and login in the same browser with the different user id. But, the application stored session cookies which only allowed to login once. Successive attempts even after logout will always log with the previous user. We found that Microsoft Windows provide an API to clear session cookies but that code needs to be called by the IE process. Since our code runs in QTP, calling that API didn't help. So we built a DLL in VC++, injected it into IE and executed that API to kill the browser session. This worked and we were able to re-use the same browser to log in with a new user id.

There are lot of challenges like these which may appear small and simple but finding a right and effective solution can be a tough task.

Alex: Have you had a chance to use custom DLLs in one of your projects?

Me: Yes.

Alex: Can you tell me a few advantages of using custom DLLs?

Me: Custom DLLs can be used with the test through DOTNetFactory or by making their methods COM visible.

Using compiled code versus VBScript has several advantages. Firstly, using compiled code means better performance. Secondly, languages such as VB.Net and C#.Net have much better and powerful error handling. The IDE of .Net products is also extremely powerful, with much better IntelliSense offered by the QTP IDE.

Object-Oriented programming languages also support Inheritance and Polymorphism, which VBScript doesn't. Most errors are made visible even before the code is compiled, which is not possible with a scripting language like VBScript. This is because scripted code is interpreted and compiled to execute only at run-time.

Programming languages also offer a wide variety of objects and techniques to accomplish tasks. VBScript is quite limited and does not support the vast range of objects supported by .Net technologies or other object-oriented languages.

Alex: Any disadvantages?

Me: Even though using custom DLLs is highly advantageous, there are still several drawbacks which are reasonable enough to not use them in automation projects.

The first and the most important of them are, not everyone in the automation team has familiarity with programming languages. This can cause a big drawback if the programmer in the team decides to leave the project, or isn't available at a time when modification to

Face to Face Interview–Round 2

the code is required. Therefore, because of an unstable DLL or a buggy DLL, the testing won't move forward.

Also, each time there is a change; the DLL must be compiled again and loaded on all machines where the DLL is being used. Even though this problem can be resolved by sharing the DLL over a network drive, having to compile the code each time can become quite troublesome.

A DLL also adds another layer between the application and QTP itself, which means, another component to manage and maintain. This can result in greater complexities in high-pressure environments.

Alex: You have mentioned in your CV that you have worked on C#. So you must have some insight into the .NET framework as well?

Me: Not completely. I have created few COM components in C# but I am not very good at the .NET concepts.

Alex: What are the two core components in .NET framework?

Me: MSIL & CLR are two main aspects of .NET framework but I am not sure if the same are the core components as well.

Alex: What is MSIL?

Me: MSIL stands for Microsoft Intermediate Language. All .NET programs when compiled produce the MSIL binaries.

Alex: What is CLR?

Me: CLR stands for Common Language Runtime. The CLR component is responsible for converting the MSIL code into the operating systemcode.

(I was shocked to see this interview moving towards C# instead of QTP, I was not prepared for this)

And I thought I knew QTP!

Alex: What is CTS?

Me: *(I know a company with that name but not sure what is it in .NET)*

I don't know.

 Note: CTS stands for Common Type System. The common type system defines how types are declared, used, and managed in the runtime. It is also an important part of the runtime's support for cross-language integration.

Alex: What is the use of partial keyword in C#?

Me: Partial keyword allows splitting a class into multiple code files within the project. This helps divide code into different sections in huge class implementations.

Alex: What is the difference between Array and ArrayList?

Me: Arrays are type specific. They can only store data of the type they are declared for. ArrayList can store object or any data type. Arrays in C# are fixed size, while ArrayList can shrink and grow.

These are the key differences I know.

Alex: What is the difference between run-time and compile time polymorphism?

Me: *(I remembered these concepts in bits and pieces and I knew if I would open my mouth here I might spell out something wrong here)*

I don't know.

Note: Compile Time Polymorphism:

Compile time polymorphism is method and operators overloading. It is also called early binding.

142   Tarun

When methods have the same name but different arguments, it is called method overloading.

In method overloading, methods perform different tasks with different input parameters.

Runtime Time Polymorphism:

Runtime time polymorphism is done using inheritance and virtual functions. Method overriding is called runtime polymorphism. It is also called late binding.

Method overriding occurs when child class declares a method that has the same type arguments as a method declared by one of its SuperClass.

When overriding a method, you change the behaviour of the method for the derived class. Overloading a method simply involves having another method with the same prototype.

In Runtime Polymorphism (just like method overloading), it is impossible to tell as to which version of the method will be executed at compile time. In other words, only at runtime do we know which method of a class will be called in case that method is overridden in another class.

Alex: What is difference between Metadata and Manifest?

Me: Metadata means data about data. Metadata is used by .NET to describe various aspects of a type, like methods supported, properties, data types etc.

Manifest is an XML file but I am not very sure as to what it is for.

> Note: The assembly manifest contains the assembly's metadata. The manifest establishes the assembly's identity, specifies the files that make up the assembly implementation, specifies the types and resources that make up the assembly, itemizes the compile-time dependencies on other assemblies, and specifies the set of permissions required for the assembly to run properly.

Alex: What is AppDomain and why is it required?

Me: The assembly we create runs in AppDomain. I am not very sure what is required for.

(I knew it was not going well now. I was not answering any of the questions)

> Note: Application domains aid security, separating applications from each other and each other's data. A single process can run several application domains, with the same level of isolation that would exist in separate processes. Running multiple applications within a single process can increase server scalability.

Alex: What is the difference between Build and Rebuild options in a solution?

Me: (finally something I knew...)

Build option means compile and link only those files which have changed since last build. Rebuild is to build each and every file regardless of whether they have changed or not.

Alex: Can we code using 2 different .NET languages in the same project?

Me: No. We need different projects for the same.

Alex: How do you install your assembly in Global Assembly Cache (GAC)?

Me: There are two ways:

- One is to drag and drop the assembly into the Windows assembly folder.
- Second option is to use the GACUTIL utility to install the assembly in GAC.

Alex: Are there any pre-requisites for installing the assembly in GAC?

Me: No.

Alex: Don't you have to sign the assembly?

Me: Yes, we do need to sign the assembly.

(I had my eyes down by now. I was nervous by now and had a bad feeling that I won't make it through this round of the interview.)

Alex: I have two assemblies with same name, same version but different public keys. Can I install them in GAC?

Me: I know that different version can be installed, but I am not sure about this one. I have never tried this. But I doubt it would work.

Alex: How do you run the IIS Server?

Me: You mean the IIS server using ASP.NET?

Alex: Yes.

Me: I have no experience in ASP.NET. I don't know how to run IIS.

Alex: What is the difference between an Interface and an Abstract class?

Me: An abstract class is a special class which cannot be instantiated. Abstract classes can have code implementation as well. Interfaces on other hand can have no code implementation. A class implementing an Interface needs to implement all methods and properties of the interface.

Alex: Any other difference?

Me: No.

> Note: Poor Nurat missed mentioning that a class can inherit many interfaces but it can only inherit one abstract class. Also a class inheriting the abstract class need not define the methods of the abstract class. It can override the methods if required.

Alex: What is the difference between readonly and const keyword in C#?

Lalwani

And I thought I knew QTP!

Me: Read-only can be modified in the class but not outside but const can never be modified. Const needs to be declared and assigned a value at compile time only.

Alex: (The interviewer stared at me and said...)

Are you sure?

Me: (I was in full panic mode by now... I stammered)

I.. I am not sure. Sir, my expertise is in QTP and not in C#. I have occasionally used C# for some tasks but I am not very well versed with all these concepts you have asked.

 Note: A const must be initialized at the time of its creation. A readonly field can be assigned to once in the class constructor allowing you to pass in the value at run-time. Declaring fields

as const protects both you and other programmers from accidentally changing the value of the field. Also note that, with const fields the compiler performs some optimization by not declaring any stack space for the field. The readonly keyword is similar to const with two exceptions. First the storage of a readonly field is the same as a regular read-write field and thus there is no performance benefit. Secondly readonly fields can be initialized in the constructor of the containing class.

A ReadOnly property can change, but it cannot be modified.

In the below example the property name will never change, but the property age can.

```
public class Student
{
  public string name
  {
  get
  {
    return "john";
  }
  }

  public int age
  {
  get
  {
    Random random = new Random();
    //random number between 12 and 18
    return random.Next(12, 18);
  }
  }
}
```

And I thought I knew QTP!

Alex: Okay, let me test your QTP skills now. What scripting languages can be used in QTP?

Me: *QTP as of now only supports VBScript.*

Alex: Are you sure QTP doesn't support any other scripting languages?

Me: *To be precise QTP scripts can only be written in VBScript language though we can initiate external scripts from within QTP which may be of different language. If we use something like*

SystemUtil.Run "cscript", "C:\MyJSFile.js" then it would run a JScript but not in its own environment. So if you consider this as QTP supporting JScript then the answer would be yes else it would be No. And in my opinion this should not be considered as JScript supported by QTP.

Alex: Fair enough. Will the below two statements have the same results?

```
SystemUtil.Run "iexplore.exe", "http://KnowledgeInbox.com/"
SystemUtil.Run "http://KnowledgeInbox.com/"
```

Me: *Well, they can have same or different output depending on system configuration. The first statement will always launch the URL in Internet Explorer.*

The second statement would launch the URL in the default browser registered on the system. If the default browser is Firefox, then the URL will be launched in Firefox.

Also there is some configuration in registry which can make the code launch the URL in explorer.exe instead of iexplore.exe. In such a case, QTP may identify the Browser as a Window object.

Alex: Is there a resolution to this explorer or iexplore issue?

Me: *Yes. There are two possibilities. One is to change the relevant key in Windows Registry. The key name is BrowseNewProcess but I don't exactly remember the path. We can easily get it by searching the key in RegEdit.*

The second option is to navigate to Tools->Options->Web->Advanced in QTP where there is an option for enabling support for Microsoft Windows Explorer.

Note: The BrowserNewProcess key is located at below path in Registry

HKEY_CURRENT_USER\Software\Microsoft\Windows\CurrentVersion\Explorer\BrowseNewProcess

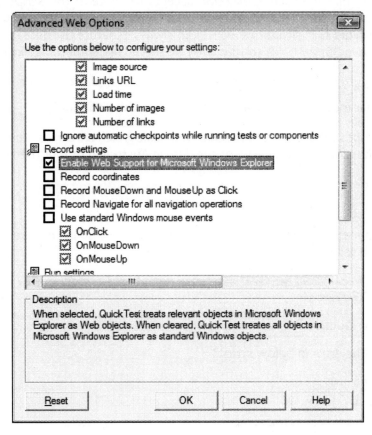

And I thought I knew QTP!

Alex: What are the potential issues in the below script?

```
Dim sDate
sDate = CStr(Date())
Browser().Page().WebEdit().Set sDate
```

Me: The code seems to have no error and it should always run fine. But a potential issue could be the format in which application accepts the date. If this script is used across different geographic locations, the format of date may change based on current regional setting. So to avoid such situations we should always generate the date in the format we want. So, if it is in DD/MM/YYYY format then we should use something like below

```
sDate = Day(Date) & "/" & Month(Date) & "/" & Year(Date)
```

Alex: Any other method you are aware of avoiding this situation?

Me: Yes. VBScript supports a method name SetLocale which we can use to set the regional settings for the script. In case we want the date in Australian format then we can use:

```
Call SetLocale("en-au")
sDate = CStr(Date)
```

The SetLocale method returns the old Local value as well. So in case we want to revert back to the old settings we can use SetLocale again.

Note: To retrieve the current Locale setting, we can use GetLocale:

```
oldLocale = GetLocale
```

Alex: What is the error in below script?

```
For i = 0 to 10
    Print i
Next
```

Face to Face Interview–Round 2

Me: (Even if I run this 100 times I won't get an error. Not sure what he is expecting...Is there something I am missing??...got it)

I forgot to ask you in which environment is this script running in? If this is running in QTP it would run perfectly fine. Also it would in QTP 9.x or higher. If executed in QTP 8.x or as VBScript outside QTP, this would error out as there is no Print statement available.

Alex: Good! I am giving you a code snippet in a QTP test:

```
Dim X
X = 10
If X = 20 Then
    Msgbox "You are inside If"
End If
```

I want you to have the MsgBox shown without making any changes to the script.

Me: (This is a strange question. Can this be done or not?)

Can we add a new Library file to this test?

Alex: No, you can't.

Me: Then I am not sure if this can be done at all.

Alex: Okay. What if I allowed you to add a Library file? How would you do it then?

Me: Then I would just add a class and initialize an object for the same and put this message box inside the class. Something like below:

```
Class Test
   Sub Class_Initialize()
      Msgbox "You are inside If"
   End Sub
End Class

Set oTest = New Test
```

And I thought I knew QTP!

This would display the message box when the test is run.

Another way is to simply add the MsgBox statement in the library and associate it with the test:

```
MsgBox "You are inside If"
```

Alex: This would work, but it is not something I expected. Any other way?

Me: Then please give me two minutes to think over this..

(He doesn't want me to use classes, he doesn't want me to modify the script...How can I do this....If only QTP had an option to change the next statement like Visual Studio this would been have been a piece of cake with a breakpoint. Breakpoint...that's it J)

This is possible. What I would do is that I would put a breakpoint on the 'If X =20 Then' line and when I am in debug mode I will go to the Watch window or the Command window and change the value of X to 20. After this I would resume the script and it would show the message box.

 Note: PowerDebug tool from KnowledgeInbox allows jumping to any valid code statement in current code context. This features helps in debugging and testing more code at run-time without stopping the test in case of errors

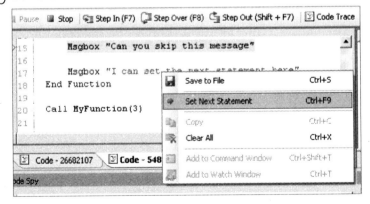

Face to Face Interview–Round 2

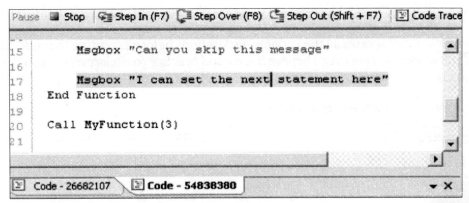

For more details refer to the below URL

KnowledgeInbox.com/products/powerdebug/

Alex: Ok. What would be the output of the below code snippet?

```
Option Explicit

X = 2
Msgbox X
```

Me: This would give an error since the variable 'X' is undefined.

Alex: What if I change the code as below?

```
Option Explicit
X = 2
Msgbox X
Dim X
```

Me: This would work as we have declared the variable.

Alex: Wouldn't this throw an error as I have used the variable before defining it?

Me: No, VBScript processes all the variable declarations before executing the code. So this would still work fine. However, I believe it is a good practice to declare all variables before they are used

Andrew: What will be the output of below script?

```
For i = 10 to 0
    Msgbox i
Next
```

Me: The loop won't run as the For loop always run in increment mode, to make it work we need to change the step value

```
For i = 10 to 0 Step -1
    Msgbox i
Next
```

Alex: Consider the below script

```
Option Explicit
ReDim arrTest(2)
arrTest(0) = "Tarun Lalwani"
```

Now I have not declared the arrTest variable with a Dim. On which line of code would I get the error? ReDim or arrTest(0)?

Me: (Another tricky one it seems!!! A ReDim should declare the variable in first place I feel)

It's a tricky question, I haven't tried this out. But from my understanding ReDim should declare the variable at first instance and the code shouldn't produce any error.

Note: ReDim is enough for declaring a variable array, we don't need additional Dim.

Alex: What is the issue in below script?

```
SystemUtil.Run "c:\Program Files\\Internet Explorer\\iexplore.exe"
```

Me: There is no issue with the code. Though we don't need to use a '\\' in the path but even if we do, there won't be a problem.

The problem will come only if we had the path starting as "C:\\". We can't use a \\ with the drive. For rest of the path, any number of continuous slashes will be treated as single slash.

Alex: Consider the below script:

```
Function Test()
    X = 2
End Function


Call Test


Msgbox X
```

What would be the output?

Me: It would give an empty message box as the value is created inside the Function Test and is destroyed when the Function scope ends.

Alex: Is it possible to fix this in a way that the value change is reflected?

And I thought I knew QTP!

Me: Yes, if we declare the variable in the global scope and change the code like below:

```
Dim X

Function Test
    X = 2
End Function

Call Test
Msgbox X
```

Then it would start working.

Alex: Is there any other way also?

Me: Any other way...The assignment needs to happen in the Global scope, so I guess we can use the ExecuteGlobal method and change the code as below:

```
Function Test
   ExecuteGlobal "X = 2"
End Function
Call Test
Msgbox X
```

This should make it work

Face to Face Interview–Round 2

Alex: What if I change the code as below?

```
X = 7

Function Test
   X = 2
End Function

Call Test

Msgbox X
```

What would be the output now?

Me: The message box now would show 2. When the first statement 'X = 7' gets executed the variable X gets created in the global scope and when we call Test Function, it uses the X from the global scope itself. And hence we get the updated value

Alex: What if I don't want the value to be updated?

Me: Then we can declare the variable X locally in the Test Function. It would make the Test Function use the local copy of X

```
X = 7

Function Test
   Dim X
   X = 2
End Function

Call Test

Msgbox X
```

And I thought I knew QTP!

Alex: How can we swap two variables?

Me: (Ahhh....that's an easy one)

```
TempVar = VarA
VarA = VarB
VarB = TempVar
```

Alex: Sorry, I forgot to tell you. No temporary variables

Me: (Huff...why did I consider myself lucky)

We can then use a Function to Swap

```
Function Swap(ByRef ValA, ByRef ValB)
    Swap = ValB
    ValB = ValA
End Function

ValA = Swap(ValA, ValB)
```

Caller: Well you have used a Function here, which gives you return value as a kind of variable only. I can modify your code as

```
Function Swap(ByRef ValA, ByRef ValB)
    Swap = ValB
    ValB = ValA
    ValA = Swap
End Function

Call Swap(ValA, ValB)
```

Face to Face Interview–Round 2

This is as good as your code earlier. This is not what I am looking for. I don't even want you to use Function also

Me: (He will never spare me. I think I will have to consider some operation on the variables

Well let me consider some values

A = 10

B = 20

Now I need A = 20 and B = 10

If I do A = A + B. I have A = 30 and B = 20

Now to change B I can have B = A – B = 10

Now A = 30 and B = 10

A = A – B = 20

Let me just check this on another values A = 5, B = 4

A = 9 then B = 9 – 4 = 5 and A = 9 – 5 = 4

So this seems to work)

We can use the addition and subtraction in this

```
VarA = VarA + VarB
VarB = VarA - VarB
VarA = VarA - VarB
```

Caller: This solution would only work for numeric values?

Me: Yes

Caller: How would you swap 2 strings then?

Me: (I was about to pull my hairs now. I didn't know if I could pull something of this off or not. I might have to do it the same way only

Finally after some struggle on the paper I was able to solve this puzzle)

```
VarA = VarB & VarA
VarB = Mid (VarA, Len(VarB) + 1)
VarA = Left (VarA, Len (VarA) - Len(VarB))
```

Alex: What is the purpose of Set statement?

Me: The Set operator is used on the left hand side of a variable and is required whenever we want to assign an object to the variable. We do not require the use of Set when the right hand side of statement is not an object. For example, consider the below three statements:

```
Set oDict = CreateObject("Scripting.Dictionary")
Set oDesc = Description.Create
Set NewClass = New MyClass
```

Alex: Have you used GetRef?

Me: Yes. The GetRef Function returns a reference object to a Function or Sub.

Alex: Can you show me a quick example?

Me: Absolutely! Consider this code:

```
Function RefFunction(sUserName, sPassword)
   MsgBox sUserName & " - " & sPassword
End Function
```

```
Set GetRefFunction = GetRef("RefFunction")
```

```
GetRefFunction "Test", "Test"
```

Once this executes, GetRefFunction will contain a reference object of RefFunction. In other words then, both the Function and its reference will perform the exact same operation.

Alex: When we declare a class in an associated library, why do we get an error using the New operator for that class in an Action?

Me: Every Action runs in its own namespace which is different from the namespace where the associated Libraries run. New operator in VBScript can only work with local classes that have the same scope as the statement where the New operator is used. The reason behind this behaviour is that, by default, Classes in VBScript have a Private scope. So, the only workaround to create a reference to an object having a different scope is byreferencing a Public object within the same scope as the class and then return the same from the Function. Below code shows a sample library Function to do the same:

```
Class myClass
End Class
```

```
Function NewMyClass()
   Set NewMyClass = New myClass
End Function
```

Now in the Action we can get the object using the Function we just declared:

```
Set oClass = NewMyClass
```

Alex: What is QFL?

Me: It is an acronym for QuickTest Function Library.

Alex: I have an application where there are two instances of the same application hosted on two different URLs. I want to write a test to compare both the websites. They have the same titles, same objects and everything. How would I work these multiple browsers and do the comparison?

Me: There are few ways I can do this

- *We can launch two browsers at the same time; identify one of them using CreationTime:=0 for the first open browser and CreationTime:=1 for the second open browser.*

- *Another way is to launch them with the application URL itself and then identify them using the OpenURL property. Since we know both applications have different URL's this would be a better approach to use.*

- *One more way is to use random numbers while launching the browser and then using the OpenTitle property for identification. In other words, we can do the following:*

```
SystemUtil.Run "iexplore.exe", "about:"& RandomNumber.Value(10000,99999)
```

Alex: What challenges do you feel you would face in doing this comparison?

Me: The task would be very challenging if during test, both browser throw Popup windows at the same location. In other words, consider the same link in both browsers that opens another browser window when clicked. Differentiating between the popup windows from both could be tough. But a workaround would be to open popup in one application, read the required information, close it, and then do the same for the second instance.

Alex: What is the Missing Resources pane in the QTP UI?

Me: It specifies what resources associated with the test are missing. We can double click on the missing resource under the Missing Resources pane, and attempt to resolve the error by locating the resource.

Face to Face Interview–Round 2

Alex: What are the types of missing resources it can show?

Me: Missing resource can be any of the following:

- Shared object repository
- Function Library
- An unmapped repository parameter
- Recovery scenario
- An Action call

These are the ones that can be found within the pane.

Alex: Aren't you still missing one?

Me: I am not sure if I missed something because these are the resources that could be associated with test. I can't imagine anything missing (I was a bit confused by now as to whether he was testing me or if I really was missing something).

Alex: Ok. There is still one missing which you can find out later.

 Note: Nurat forgot to mention the Environment variables file.

Alex: Consider a situation where a QTP script runs for long durations. I want to create an external script which can be used to call a Function inside this running script without stopping the test. How would you approach this problem?

Me: Do you mean you want to pause the test for a moment and then execute a Function and let the script continue?

Alex: Yes, you are right?

Me: Externally QTP doesn't allow accessing any of the Functions that are present in the Test. We can use QTP AOM to access some of the information but not the Test. But the possible solution I can think of is an external trigger that lets our script know that a Function needs to be executed. For this I would create a recovery scenario with trigger as a popup window. The title can be set to something like "EXT_TRIGGER.*" and then we associate a Recovery Function as the recovery action. The Function would look something like below

```
Function EXT_TRIGGER(Object)
End Function
```

The object parameter will receive the Window object when the popup trigger is found. In case we want to run a Function based on this trigger, we can use the format of trigger title as EXT_TRIGGER:FunctionName. Then in the EXT_TRIGGER Function we can extract the Function name from the title

```
Function EXT_TRIGGER(Object)
    Dim sTitle, sFunctionName
    sTitle = Object.GetROProperty("title")
    sFunctionName = Split(sTitle, ":")(1)
    Dim oFnPtr
    Set oFnPtr = GetRef(sFunctionName)
    Call oFnPtr()
End Function
```

Now the only part left is checking for this trigger in QTP. The QTP recovery settings allow us to check recovery scenario in 3 places:

- On Every step
- On Error
- Never

In our case since we want the Function to be executed any-time even if the script is running fine, we need to go for the 'On Every Step'. This may decrease the performance of the script a bit. And another issue is that if there is a big loop with no Test Objects involved then the recovery scenarios will not be checked. To give you an example

```
For i = 0 to 100000
    Print i
Next
```

If the above script is running then QTP will not catch any of the recovery triggers as the code doesn't use QTP Test Objects.

Alex: So how would you fix that?

Me: Again we have 2 options here. One is to add the Recovery.Activate statement inside the loop which forces the checking of recovery scenario.

```
For i = 0 to 100000
    Print i
    Recovery.Activate
Next
```

Second option is to add a dummy object which would get the recovery scenario fired.

```
For i = 0 to 100000
    Print i
    bFlag = Window("hwnd:=0").Exist(0)
Next
```

Alex: You said if we use recovery scenarios 'On Every Step' then the performance would suffer. How can we improve on that as well?

And I thought I knew QTP!

Me: Then other option would be to set the recovery scenario setting as 'Never'. In the scripts we can then enforce recovery scenario checking using Recovery.Activate method. Usually in case of web application it is good to override the Sync method and add Recovery.Activate with the same.

Alex: Have you worked on a project that used Captcha?

Me: Yes.

Alex: What is Captcha used for?

Me: Captcha is an acronym for Completely Automated Public Turing test to tell Computers and Humans Apart which simply means, its purpose is to prevent automation.

Alex: Why is it required to prevent automation?

Me: At times it's required to prevent certain types of software to automate tasks because their activities can be harmful to server systems. For example, if I have a website that allows comments and I do not have a mechanism that can prevent bots to post spam comments, then I will end up having hundreds of comments that bring in no value. The downside to this is that they take space on my server and I have to delete them from being seen on my website. Some bots can also circulate improper content hurting my website's reputation. The use of Captcha can prevent this abuse and help me save both time and money.

Alex: You have to Automate a user registration page which has Captcha images. How would you enter text from the Captcha image on the page?

Me: The current Captcha images are very hard to break and we cannot use Optical Character Recognition (OCR) tools for most of them. So, a few possible workarounds to automate such applications could be:

- We can ask the developer to keep a fixed Captcha image in the Test environment.

Face to Face Interview–Round 2

- *We can ask the developer to embed a hidden tag with the value of the Captcha text inside the page. We can read the text at run-time from the page and use it. This again, is for Test Environment only.*
- *Developers can provide us with an algorithm which can be used to decode the image to text.*

Alex: I have the below mentioned code in my script and a browser with the Google website open:

```
Msgbox Page("title:=.*Google.*").Exist(0)

Msgbox Page("title:=.*Google.*").WebEdit("name:=q").Exist(0)
```

Would this run fine or not?

Me: It should error out as we don't have the browser object. But we can't say for sure without running this code in QTP as I have seen weird things working in QTP.

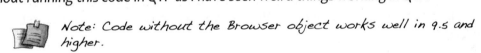
Note: Code without the Browser object works well in 9.5 and higher.

Alex: I have a script which runs fine in VBScript and I move that script to QTP. The script doesn't use any of the QTP's Test Objects; will this script work in QTP as well or not?

Me: There are two parts to every script written in VBScript language, one is the core language related code and another is host related code or objects. If we use any of the host related objects in our script, then we can't move it from one host to another. When we run a VBS file as a standalone script in windows, the default host used is WScript. So, if this object is being used in the code then the script won't work in QTP, else it would.

Alex: Can I encrypt or password-protect my code in QTP?

Me: No, we can't. There are few obfuscation tools available in market which can make the code difficult to understand but nothing except that.

Alex: Do you know any methods for hiding code from user, since we can't encrypt the code?

Me: One way to do it would create a COM based DLL and then pass all the required objects to the same and create Functions in that. Two big disadvantages I see in this approach are maintenance and scope difference. A small change required in the code would need the recompilation of the DLL. Also since we are not running code in QTP's environment, objects like Reporter, Browser etc will not be available to use and hence we cannot write the code in the same way we do in QTP.

Alex: Can we do API testing using QTP?

Me: We can, but only partially. QTP uses VBScript as its core language, so any API using data-type not supported by VBScript will not work in QTP. QTP provides an Extern utility object which can be used to declare and make API calls but the DLLs need to be C style DLLs. Any of the API that needs data type like structures, double pointers etc will not work as VBScript's data-types won't support the same. So we can only test a limited set of APIs using QTP.

Alex: What are CheckPoints?

Me: CheckPoints are verification points which can be inserted in a test to verify expected behaviour in the AUT.

Alex: Why do we need checkpoints?

Me: Without CheckPoints our scripts will never pass. However, they can fail in case an error occurs. CheckPoints are important verification points in AUT which are used to test requirements.

Alex: How many types of CheckPoints does QTP support?

Face to Face Interview–Round 2

Me: Well there is Standard, Text, Text Area, Bitmap, Image, Table, Database and XML.

Alex: There is one that you missed. Which one?

Me: Mmm....I can't remember which one I have missed.

Alex: You missed the Accessibility CheckPoint. Anyways tell me what is the difference between Standard and Text Checkpoint?

Me: A Standard CheckPoint is used for verifying a number of object properties at the same time. While a Text CheckPoint to verify text for a given Object. Standard CheckPoint can be used on any type of Test Object while Text Checkpoint doesn't work on all types of Test Objects.

Alex: How do you compare 2 images in QTP?

Me: One option is to use the Bitmap CheckPoint. But for that, one image has to be stored in the Bitmap CheckPoint and the other image has to be displayed on the screen.

Second option is when both the images are available as files; we can do a byte a byte comparison of the file. But this would actually compare the file storage format, so it may fail for 2 same images if their file storage format has even 1 bit changed.

Third option is to use a 3rd party comparison tool which can compare image pixel by pixel. One such tool is KnowledgeInbox Screen Comparison API which allows COM objects for comparing images

Note: KnowledgeInbox screen capture images can be downloaded from below link

KnowledgeInbox.com/?s=comparison+com+api

Alex: What is the difference between Image and Bitmap CheckPoints?

Me: An Image CheckPoint is for checking properties of an image like source, width, and height while a Bitmap CheckPoint checks the actual image pixel by pixel.

Lalwani

169

Alex: What are the types of tolerances you can add to a Bitmap CheckPoint?

Me: Pixel tolerance defines the count of pixels that can differ. RGB tolerance specifies the percentage by which a pixel colour can differ. These are the two tolerances supported by Bitmap CheckPoints.

Alex: What is the difference between Text and Text Area CheckPoints?

Me: Text CheckPoint is for verifying if the text is positioned properly between specified text blocks. We can specify some static text that will come before and after our text to check. This feature of Text CheckPoint enables us to test values which are dynamic in nature. E.g. 'Your order has been shipped on 21-May-2010', in this text the date would be dynamic but the part before it will always be constant.

With a TextArea CheckPoint we can check whether the given text is within specified area or region on the application. TextArea CheckPoint can be considered as a sub-set of Text CheckPoint.

Alex: What if I insert a Text CheckPoint on a WinObject which doesn't support reading the text property through GetROProperty?

Me: That won't be an issue as QTP will use OCR (Optical Character recognition) to get the text. But OCRs are not always perfect, so we might get unexpected text.

Alex: How do you check if a CheckPoint has passed or failed?

Me: The Check method that is used to execute CheckPoints returns a true/false value. We can capture that value in case we want to check for failure or success of CheckPoint:

```
bStatus = Object.Check (Checkpoint("CheckpointName"))
```

Alex: How do you disable reporting of a failed CheckPoint?

Me: We can disable the reporting through Reporter utility object

```
Reporter.Filter = rfDisableAll
```

Alex: How can you create CheckPoints at run-time?

Me: We can't create CheckPoints at run-time. They can only be inserted during design-time/recording.

Alex: How do you parameterize CheckPoints?

Me: We can map expected values to be picked from the DataTable or Environment variables.

Alex: What are the disadvantages of using QTP's built-in CheckPoints?

Me: I feel there are many disadvantages of QTP's built-in CheckPoints and that is the reason why I never use them in scripts. Few of the important ones that I know are:

- Checkpoints cannot be created at run-time
- Checkpoints are not flexible and easily portable
- When implementing complex CheckPoints like Databases and XML, the expected values cannot be easily updated at run-time.
- Updating CheckPoints during maintenance can be a tough and time-consuming task.

Alex: How can you update QTP's CheckPoint automatically?

Me: QTP provides an Update Run mode which updates all the CheckPoints with latest values from the application.

Similarly, there is a Maintenance Run mode as well which can be used for updating object descriptions.

Alex: Can we update the name of CheckPoint in QTP?

Me: Yes, we can with QTP 9.X or higher.

Alex: Does an Output CheckPoint output a value?

Me: Yes. It does but there is nothing like Output CheckPoint. Instead, it is an Output value. Output value can output a value to Global or Local DataTables or even to an Environment variable.

Alex: What is the difference between Checkpoints and Output Value?

Me: CheckPoints are for verification purposes whereas Output values are for capturing actual values from the AUT. Using an output value, we extract different property values into DataTable or Environment variables. We can then use these values in our script.

Alex: Can Output value fail a Test?

Me: Yes, but only if the object which is used for outputting the value doesn't exists. There is no validation in case of output values.

Alex: What is the difference between Window and Dialog Test Objects?

Me: I doubt there is much difference between them. Dialog is specific type of a Window which has a Window style set as a Dialog. We can identify a Dialog object using a Window object as well with same description properties. The only other difference I know is that Dialog object doesn't support the RunAnalog method while the Window object does.

Alex: How do you capture a screenshot when any error occurs in the script?

Me: The easiest way would be to create a recovery scenario which triggers on Any Error and call Function as recovery action. In this function we can insert a Desktop.CaptureBitmap to capture all the objects currently displayed on the Desktop.

 Note: We can also use QTP settings to capture screenshots in case of errors

```
Setting("SnapshotReportMode") = 1
```

0 - always captures images.

1 - captures images only if an error occurs on the page.

2 - captures images if an error or warning occurs on the page.

3 - never captures images

Alex: How do you capture a screenshot only when you reported a failure using Repoter.ReportEvent?

Me: We can create another Function called ReportFail which takes 2 parameters and hard-code micFail. The 2 parameters passed to the Function can be used with Reporter.ReportEvent StepName and Details. In the same Function we can capture the screenshot.

Alex: Your method would require too much maintenance. Any other option which requires less maintenance?

Me: We can create a Function ReporterReportEvent and takes 3 parameters. This way will only have to replace the "." in our libraries from Reporter.ReportEvent

Alex: This is better from your previous method of course, but now let me put a restriction that the existing code should not be changed. Can this still be done?

Me: I am not sure but let me just think

(I am not sure if there is an answer to this or he just wants me to say no affirmatively. I need to run my brain on full throttle to get something for this one fast. Let's run a logical test

```
Reporter.ReportEvent micFail , "X", "Y"
```

I cannot override the ReportEvent method of Reporter object directly.

micFail is a constant so I can't do anything with that.

The other 2 parameters are constant, so, again no use.

If only it was ReportEvent method without any object I would have just added a new method.

Let me get back these views to him atleast)

Well if we look at the micFail statements it would look like

```
Reporter.ReportEvent micFail , "X", "Y"
```

Now only if the ReportEvent method was a direct built-in method instead of the Reporter object I could have overridden it by defining a new Function but with this, it doesn't seem possible.

Alex: Well your thought is very near to the actual answer, think hard.

Me: (I have told him only about Function, so it has to be something related to Functions, but what? Oh yes, what a stupid mistake)

I think I got it. What we can do is whenever we report a failure; we always use micFail constant which would have some value. I am not sure what exact value it has but we can easily get it through a print or MsgBox statement. Now we can create a new Function with name micFail and return this constant value:

```
Function micFail()
    micFail = 1 'Assuming 1 is default value of micFail
    'Capture the screenshot when error occurs
    Desktop.CaptureBitmap "<PATH>"
End Function
```

So whenever in our script a statement like below gets called

Face to Face Interview–Round 2

```
Reporter.ReportEvent micFail, "X", "Y"
```

micFail will actually call the Function we created and hence the screenshot will be captured.

(phew! That was cool, but I guess couldn't have figured out without his hint)

Alex: How would you launch a notepad without any code to your script?

Me: We can add it in the Record and Run Settings and specify QTP to launch it. This way when QTP runs the script, it will automatically launch the notepad application.

Alex: What Add-ins do we need to test a Mainframe application?

Me: Mainframe applications are used mostly through a Terminal Emulator (TE). We need the TE Add-in to automate such applications. But TE Add-in doesn't support all emulators; we first should check the TE Readme to check for the supported emulators. In such cases we can identify the emulator as Window object and use methods like GetVisibleText, Type to do limited operations on the emulator

Alex: Can we do it without TE Add-in?

Me: We can but for that we will have to use the Type method and GetVisibleText for screen scrapping. It is possible to automate using this approach but it increases the effort in coding and also may not be a robust solution.

Alex: What is the maximum size of an array that we can declare?

Me: Every language has some variable memory heap and some maximum size possible of that heap. So as long as VBScript has enough memory available, we should be able to declare an error of any size. Also, the array size shouldn't violate any other limitation of VBScript.

Alex: Other limits like?

Me: Array indices are long data type, so the size has to be within the max positive value of a long data type.

And I thought I knew QTP!

> Note: Long data type can take values from -2147483648 to 2147483647

Alex: What is the maximum value of a Long data type?

Me: It would be 2^31 – 1 but I don't know what exactly it evaluates to.

Alex: Aren't Long data type 32 bits? So why have you use 31 here?

Me: The first bit is the sign bit as VBScript doesn't support any unsigned data types.

Alex: I have launched object spy and clicked on the hand button. Now the object I want to spy is not in the active window, what can I do to Spy the object?

Me: We can press the ALT + CTRL button to disable the spy mode temporarily and then switch between windows using ALT + TAB buttons. Once we have the required window in focus we can press the CTRL key again to activate the spy mode. Similarly we can use CTRL key only to disable spy mode for doing mouse clicks on the application before spying on the object.

Alex: Can we use the Object Spy without launching QTP?

Me: No, the Object Spy is only available through the QTP UI.

Alex: How do you simulate a keyboard typing on to a WebEdit?

Me: One way is to change the ReplayType settings for the Web environment:

```
Setting.WebPackage("ReplayType") = 2 'Mouse
Object.WebEdit("test").Set "Typing text"
```

Once we change ReplayType to 2 QTP will type on the text instead of setting the text directly.

Another way would be to set the focus on to the object and then type on the Browser window itself:

```
Hwnd = Browser().object.HWND
Object.WebEdit("test").Click
Window("hwnd:=" & Hwnd).Type "Typing text"
```

Alex: What are QTP Environment variables?

Me: QTP provides a utility object named Environment. This object allows creating variables which are shared across Actions and Libraries.

Alex: How many types of Environment variable are there?

Me: 3 types – Built-in, Internal and External. Built-in ones are those for which QTP updates the values automatically. They provide information like OS, UserName, Current Test directory, current iteration, current Action name etc. Internal variables are those which we define at design-time and external are those which we load from an external file.

Alex: Any other difference between internal and external Environment variables?

Me: Yes, an External variable's value is read-only and cannot be updated at run-time.

Alex: How do you load Environment variables at run-time?

Me: Environment object provides a method named LoadFromFile which can be used. This loads the variables as external variables and they can't be updated at run-time.

Alex: How do you check if an Environment variable exists or not?

Me: There is no method provided by the Environment object for this. The only way to do this is to try access the environment variable and check if any error was raised. If an error gets raised then we can assume the variable doesn't exist:

```
On Error Resume Next
  bExists = True
  Err.Clear
  val = Environment(key)
  If Err.Number then bExists = False
On Error Goto 0
```

Alex: How can we export Environment variables back to XML?

Me: QTP doesn't provide any way to enumerate if all variables exist and also no method for exporting it back to XML. The only way to export would be to create a Function that can write data to XML. However, the Function would require the names of all the Environment variables which need to be written to the XML file.

Alex: You said LoadFromFile method will load environment variables as read-only, What if I want these variables to be read-write. How can I do that?

Me: In that case we will have to read the XML file ourselves using our custom code and then loop through each variable and define the environment at run-time. All Environment variables defined in this manner would not be read-only at run-time.

Alex: Why or rather where do we use Environment variables?

Me: There are no defined guidelines per se. Different users use them in different manner. Some use it to pass information from one Action to the other. I usually use it only for Application or Framework configuration. E.g. if there is an application URL, I would use that as an Environment variable. Another approach I always follow is that, I never call Environment object directly rather I always call a supporting Function (like GetEnvironment) which then returns the value of the Environment variable.

Alex: What is the use of this convention when you are directly calling Environment object only?

Face to Face Interview–Round 2

Me: *I generally take this approach keeping in mind of any future enhancements that my framework would require. For example, at times, the client has requirements where, a provision to override the URL with different URL for certain test cases is necessary. Now in such cases, it is not always possible to keep on updating the Environment XML and then execute these test cases. So, in such a case, I just enhance the Function to check for the variable in DataTable first and if it is present then the DataTable value takes preference. In other words, for the test case where a different URL is required, we just add the parameter name to the DataTable and update the DataTable cell with a new URL value. This adds plenty of flexibility in terms of customization and eases maintenance.*

Alex: What is an Object Repository?

Me: *Object Repository allows storing a Test Object's logical name and corresponding identification properties. It provides a means of storing definitions that allow execution of scripts against object definitions that we would generally expect to map at run-time. In simple terms then, an Object Repository facilitates a mechanism that enables QTP to find objects in the AUT by referring to object descriptions in the Object Repository.*

Alex: What are the types of Object Repositories?

Me: *QTP supports 2 types of Object Repositories. Local Object Repository is automatically created for every Action whereas a Shared Object Repository can be associated with any of the Actions in the test.*

Alex: What is the difference between Local OR and Shared OR?

Me: *Objects stored in a Local OR take preference over objects stored in a Shared OR. So, if the object is defined in Local OR as well as Shared OR, then QTP will always pick the definition from Local OR. Also, both repositories are stored with different extensions. Local ORs are stored as* `.bdb` *files and Shared ORs arestored as* `.tsr` *files.*

Alex: How can we convert Local OR to Shared OR?

Me: There is no tool available for this as such. But we can open the Object Repository of the Test and then export the Local OR as a Shared OR. In other words, this is more of an export process than a conversion process.

Alex: Which is better to use: Local OR or Shared OR?

Me: In my opinion, Local ORs are useful when the application is small and also the scripts are less in number. Else, Shared OR is suggested.

Also, some people argue that having many reusable Actions can completely ignore the need of creating Shared Object Repositories as each of the objects are stored in Local Repositories, just in different Actions. However, in my opinion, this is more of a personal way of working. I feel this approach would take more creation and maintenance time as opposed to using Shared OR.

Alex: What are challenges faced when we use a Shared OR?

Me: There are lot many challenges and that seems mostly because of a poor design behind the OR concept

- QTP stores OR in binary format which makes it impossible for others to create tools that can help make OR creation easier

- Shared OR can be shared across Tests. But it can't be edited by multiple users at the same time. Shared OR are not like shared excel files where multiple users can update the file at the same time

- This is a huge bottleneck when the team's size is big

- Any wrong change in Shared OR can impact all the Test cases and there is no trace of the changes done. It becomes difficult to track if that object was changed by some other user by mistake or intentionally.

- There is nothing like partial load. As the OR size grows huge and QC integration is also used, the file download delay has significant impact on script initiation time

- In case we create multiple Shared OR per user and want to merge them into one, then QTP only allows merging 2 files at once. Also the merged OR has to be shared to a new file. This makes the merging process very cumbersome and practically unrealistic to implement

- Shared ORs cannot be associated with Library files. We always need to utilize an Action for associating Shared ORs. If I have some code which needs to be executed when associated Libraries are loaded, then I can't use Shared ORs as there is no Action loaded yet. This means the Shared OR is unavailable as the Shared OR only maps to Actions.

So there are lot many challenges when you use Shared OR

(ahhh...it's not a easy job as it seems J)

Alex: You listed so many disadvantages of Shared ORs, so how do you overcome such issues in your project?

Me: Well in some of the projects, we have taken that hit as the client wants the framework to be built with ORs. In projects where feasible, we implement a Descriptive Programming based OR.

Alex: Can you elaborate more on your DP based OR?

Me: Well there are different types of DP OR than can be implemented. One is just through VBS files where we declare global variables and assign them the descriptions

```
oBrw = "title:=MyApp"
oPg = "micclass:=Page"
oTxtLogin = "name:=txt_Login"
```

In the script we can directly use the variables like this:

```
Broswer(oBrw).Page(oPg).WebEdit(oTxtLogin).Set"user"
```

Then another approach we can follow is to create an Excel Worksheet of keywords to map object definitions

Keyword	Object Definition
oBrw	Browser("title:=Browser")
oPg	oBrw;Page("micclass:=Page")
oTxtLogin	oPg;WebEdit("name:=txt_Login")

To utilize these keywords, in the script we load them into a Dictionary object and create a GetORObject Function. In the GetORObject Function, we take in the keyword for the object and then resolve all dependencies of the parent object and return the complete string definition of the object. Once we have the string definition, Eval statement can be used to get the object for the same.

Since we use Excel to store these keywords, they can be easily shared across multiple users and if one requires, even a database can be used in a similar fashion. Utilizing a Database makes it possible to create multiple updates at the same time is possible and we can add more flexibility to this DP based OR if there is such a requirement.

Alex: So your final words on DP versus OR is that DP is better?

Me: Well, not really... Once we move to DP there are a good number of features or rather convenience we lose. OR object identification can be a bit faster compared to DP. We lose IntelliSense because it is only available for objects stored in the OR. This increases difficulties in development as we need to refer the object names from Excel and as the number of objects grow, this look-up can add errors and increase script creation time. Also, as we fill in this DP Excel manually, it increases the chances of human errors and time to setup the Object Repository. So, DP based ORs also has disadvantages. I usually decide between DP or OR based on client's requirement as well as their technical competency for future maintenance.

Face to Face Interview–Round 2

Alex: How do you define a constant at run-time?

Me: We can use the execute statement to define a constant at run-time

```
X = 20
Execute "Const cX = " & X
```

Alex: Can we store an object in a constant?

Me: No, a constant can only have values.

Alex: I want to define an object constant then what are my options?

Me: One way could be to use a private variable and then create a Function to return it:

```
Private myObject
Function ConstObject()
   Set ConstObject = myObject
End Function
```

This way ConstObject can be used for getting the object but there is nothing to set it.

Alex: But I can still access the Private variable?

Me: We can encapsulate that in the class and fix that issue as well.

Alex: How will you do so?

Me: Consider the following code:

```
Class MyClass
    Private myObject
```

And I thought I knew QTP!

```
    Public Function ConstObject()
  ConstObject = myObject
  End Function
End Class
```

In the class, the variable myObject cannot be changed from outside.

Alex: What string Functions have you used?

Me: Almost all of them

Alex: Can you name all of them?

Me: (if you insist so much....)

Left, Mid, Right, Split, Join......mmm....LTrim, RTrim, Trim, Len, InStr, InStrRev, Replace, StrReverese, String, StrComp

Alex: I have a main string and a sub-string. I want to find how many occurrences the sub-string has in the main string. Can you write some code for that?

Me: We can run a loop using InStr for this

```
Dim iCount, iFound
iCount = 0
iFound = 0
iFound = InStr(iFound + 1, mainStr, subStr)
While iFound <> 0
   iCount = iCount + 1
```

Tarun

```
iFound = InStr(iFound + 1, mainStr, subStr)
Wend
```

Alex: This looks good, can you optimize this?

Me: (What???? It's just a simple loop, what more optimization does he want?)

Sir, this is a solution using one loop only, what kind of optimization are you talking about?

Alex: Let me re-word my statement; can you do it without a loop?

Me: If we want to do it without a loop then we will have to use some built-in Function.

(Which one...which one...i guess Split could do the job)

We can use a split and retriee the number of elements in the array:

```
iCount = UBound(Split(mainStr,subStr))
```

If there are 2 occurrences of sub string then array would have 3 elements and upper bound as 2, hence UBound will give us the count.

(it was easier than I had thought....)

Alex: There is one more way and that is to do with the Replace function. Can you figure it out?

Me: (Replace would remove all occurrences, so length would go down)

Yes, even replace would work.

```
iCount = (Len(mainStr) - Len(Replace(mainStr, subStr, "")))/Len(subStr)
```

Alex: Can we have optional parameters in a VBScript Function?

And I thought I knew QTP!

Me: We can but only partially. This approach is not very common though. Consider the below Function:

```
Function TestParams(A,B,C)

End Function


Call TestParams(1, 2, 3)
Call TestParams(, 2, 3)
Call TestParams(, , 3)
Call TestParams(1, , 3)


Function TestParams(A,B,C)
   If VarType(A) = vbError Then
      'default value of the parameter has not been passed
   End If
End Function
```

But total arguments to be passed to the Function still have to be intact.

(I could see the look on his face, he didn't know about this for sure)

Alex: If I give you a logical name of an object present in Object repository, how would you get the class of an object?

Me: QTP doesn't provide means of achieving this directly, so we will have to use workarounds. One way is to use various Test Object types and see which one doesn't give an error. But for this, the object needs to be the topmost object in the hierarchy. E.g.

```
Function GetObjectClass(ByVal LogicalName)
    GetObjectClass = ""
    Dim arrTypes
    arrTypes = Array("Browser", "Window", "Dialog", "JavaWindow")
    On Error Resume Next
    For each sType in arrTypes
        Err.Clear
        Set TempObj = Eval(sType & "(" & LogicalName & ")")
        If Err.Number = 0 Then
        'Match found
        GetObjectClass = sType
        On Error Goto 0
        Exit Function
        End If
    Next
    On Error Goto 0
End Function
```

Another possibility is to extract this information from the Object Repository using its XML. For this, the pre-requisite would be that the OR should be available in XML format and the object's logical name also should be unique.

Alex: How can you get the name of the Function executing currently?

Me: If we are in debug mode, then the Context list box in the Watch or Variable window will have the current Function

And I thought I knew QTP!

Caller: And how about fetching that information in code?

Me: *For that we can always define a constant inside the Function with its name. This way we will be able to access this information inside the Function*

```
Function CurrentFunction()
    Const FUNC_NAME = "CurrentFunction"
End Function
```

In any case we will have to write this information somewhere through code and latter access is the way we want it. QTP/VBScript doesn't provide anything to access this information

 Note: PowerDebug tool from KnowledgeInbox adds capabilities to access current Function name, caller Function name, caller trace etc.

```
Function ToBeCalled()
    'Load the information regarding current function
    PowerDebug.LoadInformation()
    Print("Function=" & PowerDebug.FunctionName)
    Print("Caller=" & PowerDebug.Caller)
    Print("Stack Trace=" & PowerDebug.StackTrace)
    Print("Current Code=" & PowerDebug.CurrentCode)
End Function

Function CallToBeCalled()
    Call ToBeCalled()
End Function

Call CallToBeCalled
```

Face to Face Interview–Round 2

For more details refer to the below URL

KnowledgeInbox.com/products/powerdebug/

Alex: How do you check that max length of a given WebEdit is 5?

Me: We can check the maxlength using `GetROProperty` and check if it is 5 or not. The property that determines the number of characters that can be entered in the WebEdit object is called `maxlength`.

Alex: What if there is no max length set at all but the textbox still can't accept a value greater than 5?

Me: If max length is not set then the textbox may be handling keyboard events in restricting the length.

To test this, we clear out the value of textbox using `.Set ""` and then we can use SendKeys to send a text of length greater than 5. Since the WebEdit won't accept anything above 5 chars we can read the value of the WebEdit again and check how many characters did it accept and they should be the first 5 characters of the text we sent using SendKeys.

Alex: Please look at this code and tell me what do you observe?

```
Dim X
Set X = New MyClass
Dim Y
Set Y = X
Dim Z
Z = X
```

Me: (I thought to myself that MyClass is not there in the code, does he want to check on that or report the error on Z = X?)

Lalwani

And I thought I knew QTP!

Is MyClass available in the scope of the code?

Alex: Yes it is.

Me: (Ok, now it has to be Z = X)

Then there would be an error at the line Z = X. We haven't used the Set operator for the same.

Alex: Is it possible that we have the same code and not get the error?

Me: (I am not sure if he is trying to test me or there is something I am missing)

I don't think so. It will always give an error.

Alex: Okay, let me tell you that you can't predict that this would pass or fail. Now, you can tell me would it pass?

Me: (Phew!!! I was still not getting anything on this. MyClass is a class so its reference will always be an object. I was still 100% sure that it would not pass)

I can't think of anything where this code would pass.

Alex: Okay.

 Note: Nurat missed a very tricky part in this question. The interviewer hadn't given any code related to the MyClass. It is MyClass definition that could alter the result of this code. Consider the below example:

```
Class MyClass
   Public Default Property Get Value()
      Value = 2
   End Property
End Class
```

Face to Face Interview–Round 2

```
Dim X

'X has the object reference to MyClass
Set X = New MyClass

Dim Y

'Y has a reference to MyClass object stored in X
Set Y = X

Dim Z

'The default Get Property is called on the MyClass object and
value property is accessed
'The below code is equivalent to Z = X.Value
Z = X
```

 Note: Default keyword is only allowed on one of the Get Property method of the class in VBScript.

Alex: Does QTP offer a Goto statement?

Me: No, it doesn't and it is a limitation of VBScript language.

 Note: QTP/VBScript doesn't provide a goto statement. But PowerDebug tool from KnowledgeInbox adds these capabilities to QTP

```
'Text to be present before the tag
PowerDebug.GotoPrefix = "':"

'Text to be present after the tag
PowerDebug.GotoPostfix = ":"
```

```
'Jump tag without prefix and post fix
PowerDebug.Goto "JUMPLOCATION_NEW"
Msgbox "This message should not come because of above goto statement"


'Jump tag with prefix and postfix
':JUMPLOCATION_NEW:
Msgbox "You are here after a goto"
```

For more details on the product please refer to the below URL

KnowledgeInbox.com/products/powerdebug/

Alex: What is the maximum number of lines of code you can have for a framework?

Me: Number of lines would not make any difference. It has no relation with the concepts behind a type of framework. We can have as many lines as we want.

Alex: How can you retrieve the Native property of a Web object?

Me: We can use 'dot object' (.Object) on the Test Object and then access the native property.

Alex: What if I don't want to use the '.Object' property?

Me: Then we can use GetROProperty, but it only works for a limited set of properties

```
GetROProperty("attribute/<name>")
```

Also, a limitation with this approach is that we cannot use it for Browser or Page objects.

Alex: What is the issue with the code below and how would you fix it?

```
Set objSearch = Browser("Google").WebButton("GoogleSearch")
Set objText = Browser("Google").WebEdit("q")
objText.Set "Test 1"
objSearch.Click
<Synchronization occurs here>
objText.Set "Test 2"
objSearch.Click
```

Me: The issue is the re-use of `objText` *and* `objSearch`*. When we perform any operation on an object, through its reference, it creates a hard reference to the object. If due to a change in application's state, the hard reference goes out of scope. In our case, the application state changes when we click the Search button on line 4. So, before reusing the object we need to re-instantiate it. This can be done by using RefreshObject method on the Test Object in case of QTP 10 or higher. For lower versions of QTP, it can be done by using the Init method.*

Alex: How does QTP internally identify an object? I am not talking about mandatory, assistive and ordinal identifiers over here. I am more interested in how does it internally do it?

And I thought I knew QTP!

Me: (I guess if I had known that in such detail, wouldn't I have created a competitive Automation software against theirs.☺)

The only thing I can say regarding this is that every technology is different in its internal working and that is why QTP has different Add-ins for each. Each Add-in uses specific techniques for object identification. For example, Standard Windows objects are identified using Windows API, web objects using DOM, .NET and JAVA through hooks, Siebel through CASCOM APIs etc. But these are simply my logical guesses; how it is internally implemented may not be close to this.

Alex: Can a private Function in a Library file be viewed or accessed by the Test?

Me: No. A private Function in Library file can only be accessed by any of the associated Libraries but not by any of the Actions.

Alex: I have 3 libraries associated with my Test – Lib1, Lib2 and Lib3. Lib1 is at the top in the order and Lib 3 is at the bottom

Lib1 has below code:

```
Msgbox "Lib1"
```

Lib2 has below code:

```
Option Explicit On
X = 2
```

Lib3 has below code:

```
Msgbox X
```

When I run this test what would be the output of execution of the all the associate Libraries?

Me: To load all the Libraries, QTP combines them by adding the bottom-most file first and the top-most file at the very end. So, the combined code would look as:

```
Msgbox X
Option Explicit
X = 2
Msgbox "Lib1"
```

Now, QTP while creating this combined library removes Option Explicit from all the libraries. Only if the bottom library has Option Explicit then it is used in the final global library at the top. In this case since the middle library has Option Explicit so there would be no impact. The output of this would be first a empty message box and then Lib1.

Alex: What is the output of below code? Or would there be an error?

```
Function TestMe(A)
   TestMe = A + 10
End Function
Function TestMe(ByVal A)
   TestMe = A + 20
End Function

Print TestMe(20)
```

Me: There will be no error and the output would be 40. VBscript allows re-defining a Function and it uses the last or latest one loaded.

Alex: Have you worked with the Dictionary object?

Me: Yes, they are used to store key-value pairs. The value can be looked up using the key.

Alex: What will be the output of the below code?

```
Set oDict = CreateObject("Scripting.Dictionary")
oDict.Add "IN", "INDIA"
sCountryName = oDict("in")
Print "Dictionary count - " & oDict.count
Print "Dictionary Value for in - " & sCountryName
```

Me: The output would be as

Dictionary count – 2

Dictionary Value for in –

Alex: Can you explain why?

Me: We have used a different key "in' instead of "IN". Even though they read the same characters, one is lower-case whereas the other is upper-case. What I am trying to say is that, the keys are case-sensitive and the Dictionary object has a feature that creates a new key and assigns it an Empty value if a key that is accessed doesn't exist in the dictionary.. That is why the count increases by 1 and a blank value is given.

Alex: What needs to be changed in the code to fix this?

Me: We need to add a small line of code to make the keys case in-sensitive:

```
Set oDict = CreateObject("Scripting.Dictionary")
oDict.CompareMode = vbTextCompare
```

This needs to be done before adding any keys to the dictionary

Alex: What are user-defined Functions in QTP?

Face to Face Interview–Round 2

Me: There are few Functions provided by QTP, few by VBScript and then our own Functions. So, all Functions written by us as users are called user-defined Functions.

Alex: No, I was not talking about that. Have you heard of RegisterUserFunc?

Me: Yes, RegisterUserFunc allows us to override existing Test Object methods or add new custom methods to the Test Objects.

Alex: How would you use the same to set all values as Upper case values in WebEdit?

Me: We will create a new `Set` *Function and then in that update the value to upper-case before setting it in the target field:*

```
Function NewSet(Obj, Text)
   Obj.Set UCase(Text)
End Function

RegisterUserFunc "WebEdit", "Set", "NewSet"
```

Now, whenever we use the QTP's Set method, it would actually call our NewSet method automatically.

Alex: Can we add a new method to the Reporter object using RegisterUserFunc?

Me: No, RegisterUserFunc is used only for QTP's Test Objects and not for utility or other objects.

Alex: I have a Script A which has Action A and Library A. I have another Script B with Action B and Library B. Now I insert a call to Action A in Action B. Action A is re-usable in Script A. Would this call work or not?

Me: Yes, the call would always work. But Action A will execute successfully or not will depend on its dependency on Library A. When we call external Actions, our test doesn't inherit the Library files of the external Test, so if our external Action uses Functions from Library A and these Functions are not available in Library B then there would be a failure.

Alex: Have you worked with DotNetFactory?

Me: Yes, it was first introduced in QTP 9.2. It is a very handy utility object that allows creating and accessing .NET class objects from QTP.

 Note: DotNetFactory is available in QTP 9.1 as well.

Alex: How do you use it?

Me: The DotNetFactory object provides a method named `CreateInstance` which takes a maximum of 3 parameters. First, is the TypeName, second is the Assembly name or location and lastly, constructor parameters could be included, if any.

Alex: Where would you actually use this?

Me: There are different uses for this. One could be using default classes of .NET like arrays, stack and queues etc. This saves us from writing lot of code for supporting such data structures in QTP.

Another advantage is when the task to be done cannot be done through QTP, but needs a powerful language like .NET. In such a case we can create a DLL using any of the .NET languages and instantiate the object using DotNetFactory. This way can call the method in the DLL and execute complex tasks which might not have been possible in QTP. In other words, to use custom .NET DLLs, we do not need to register the assembly with the Windows registry. Instead, we can use `CreateInstance` to bind to the assembly directly within QTP without using utilities such as Regasm.exe to register the DLL globally.

Alex: What are the limitations of DotNetFactory?

Me: DotNetFactory is based on reflection technique which searches the method in assembly when a call is made. This means some overhead is involved in each method we call on the object. So using a .NET object in a big loop can hamper script performance.

Another limitation is that there is no way to call static method on a class. DotNetFactory only works with objects of classes.

One limitation that is more related to QTP IDE rather than DotNetFactory is IntelliSense. QTP 10 and higher versions provide IntelliSense for objects created using CreateObject but not for DotNetFactory.

Alex: What could be the issue with below code

```
Browser("Yahoo").Navigate "www.google.com"
Browser("Yahoo").Sync
```

Me: I don't see any issues as Yahoo is just a logical name and we still test the same thing on the Google website as well. The only problem would occur when we are identifying the browser using its title and not using CreationTime.

Alex: Consider I have an application with 10 different pages. Using Object Repository, can I have the same Browser and Page objects for all objects under those 10 pages?

Me: Yes. It's possible using any of these approaches:

- *Browser uses CreationTime and Page uses a Regular Expression: .**
- *Both Browser and Page objects use a Regular Expression*
- *Browser uses CreationTime and all objects are moved under Browser. In this case, the Page would not be considered to build the object hierarchy.*
- *Browser uses a Regular Expression and all objects are moved under the Browser.*

Alex: I want to run an Action for non-continuous rows instead of a range, how can I do it?

Me: There are two ways I can think of right now to do it

- One is to add a parameter to the DataTable and have a Y/N flag set in it. After checking the flag we can use ExitActionIteration, if required.

- Another way would be to take the iterations required in any array and run a loop. Inside the loop, we can call the RunAction method and run a single iteration of the Action for a specific row.

Alex: How do I find Row number of a data in DataTable?

Me: Unfortunately, QTP doesn't provide any find method for DataTable so we need to use a loop and then find the value. The below code would search for a value in all Rows for ColumnA:

```
For i = 1 to DataTable.GetRowCount
    If DataTable.GlobalSheet.GetParameter("ColumnA").ValueByRow(i) = "SearchValue" Then
        Msgbox "Value found"
        Exit For
    End If
Next
```

Alex: How do you connect to databases in QTP?

Me: It is the same way we do it in VBScript. We create an object of 'ADODB.Connection' using CreateObject and then opena connection string which is specific to the database you are connecting to do.

Alex: Can you tell me how you will connect to an Oracle database?

Me: I don't remember the connection string format but I usually use 2 methods to find the same:

- Refer to **www.connectionstrings.com** for connecting string format
- Or I create a UDL file on my desktop and configure it to connect to the database and once the connection succeeds, I just open the UDL in notepad to get the string.

Alex: Can we compare two databases in QTP?

Me: Yes we can. But we would require a few Functions which can compare two records in the RecordSet. Also, the complexity would increase in case we don't have primary or unique keys in the table to identify the record. This is because if I have 4 records in one table and 3 in the other, then just saying that there is a mismatch wouldn't help much in report. So we need to take one record from source table and use the primary or unique key combination to get the same details from the destination table and then do the comparison.

But another thing I would like to highlight here is that we don't need QTP for this, we can just do this with VBScript itself.

Alex: What is a Regular Expression?

Me: Regular Expressions are special characters which can be used for matching a pattern with actual string.

Alex: What Regular Expression would you use to match a numeric value?

Me: There are two options we can use:

One is to consider any digit between 0-9: [0-9]+

And other is to use special pattern character for digit: \d+

And I thought I knew QTP!

Alex: Give me an example where and why would you use Regular Expression in QTP?

Me: Regular Expressions can be used for different purposes. One is while identifying an object. Consider that we have a Window which has title "My Test App – ag8898" and the ag8898 is the node name of the current select node from a tree in the application, which is a dynamic value. If we navigate to a different node in the tree, the title of the window would change. In such a case, instead of identifying the window with the full title, we can use a regular expression instead of dynamic text. The pattern in our case would be "My Test App - ." where dot means any character and star means any number of occurrences of the previous character.*

Another place where we can use regular expression is to extract or clean up data. Consider I have string "1, 2a2b3 cr" and I want to extract only the numbers from it. In such a case I can use a pattern which represents not a digit pattern "[^\d]+" and then replace this with a blank string

```
strText = "1, 2a2b3 cr"
Set oReg = new RegExp
oReg.Pattern = "[^\d]+"
oReg.Global = True
Msgbox oReg.Replace(strText, "")
```

Alex: What is Smart Identification?

Me: Smart Identification is a feature of QTP which is used when QTP is not able to identify an object during run-time or during highlight.

Alex: You mention highlight. What exactly do you mean by that?

Me: I meant when you highlight an object from OR, QTP can also use Smart Identification in case it is enabled.

Alex: Ok, continue…

Face to Face Interview–Round 2

Me: Smart Identification uses two set of properties: one is the base filter set and another is optional filter set. When QTP fails to identify the object, it activates the Smart Identification algorithm, which takes all the properties in the base filter and checks for number of matches. If there is no unique match, then QTP picks properties one by one from the optional filters until it finds a unique match.

Alex: If I add an object to the OR and I keep only 2 properties for identification. How does QTP know what values it needs to use for other properties during Smart Identification?

Me: When we add an object to the OR, QTP will remember values of all other properties as well. They are kept hidden in the Object Repository and can be seen when we export the repository to the XML format. That is the reason why Smart Identification can only work on objects present in OR and not using Descriptive Programming.

Alex: Smart Identification can let us run tests even when the objects have changed slightly. So to make our test robust, we should always have Smart Identification enabled. What are your views on this?

Me: I would never want to hit my target with my eyes blindfolded. I may miss to notice that I just got lucky or I may also miss what went wrong when I don't hit the target. Having Smart Identification is just like blind folding yourself from problems that need be taken care of. Personally, I am totally against using this feature as it could lead to issues which may be hard to debug. It can also create erroneous results in scripts where an object is reported to exist even when it doesn't exist.

Alex: How do you disable Smart Identification during run-time?

Me: There are two possible ways. One is to use QTP AOM to disable it. Second is to use the Setting object to update the value of DisableReplayUsingAlgorithm as 1.

Note: You can use below code to disable Smart Identification

```
Setting("DisableReplayUsingAlgorithm") = 1
```

```
'or using AOM code

CreateObject("QuickTest.Application").Test.Settings.Run.
DisableSmartIdentification = True
```

Alex: Consider you have to automate 15000 Manual Test Cases in a week. What would you do?

Me: One thing in Automation is that based on the functionality we might only have 100 scripts to support 15000 manual Test Cases. The only thing that would change is the data being fed to these scripts. So, automating 15000 Test Cases would only be possible if the number of scripts to be developed is small. Given such a challenge, my first priority would be to analyze the number of unique scripts that would be required. Once that count is available we can prioritize which of the scripts need to be developed first. Though, I will assume to complete such a task we would already have a framework ready and a task force which is capable of doing things fast.

Alex: How do you test PDF files with QTP?

Me: QTP has no direct support or API for testing PDF files. I have not come across any solution which allows testing PDFs very well. The ways we can test PDFs are

- Convert the PDF to text using a converter and then read the text file to check for necessary information

- Use Acrobat SDK APIs to extract data from the PDF. But this gives very raw data and hence is not very useful.

There are few 3rd Party COM based libraries also for editing but none of them are free

Alex: Does QTP support testing Siebel applications?

Me: Yes, QTP supports Siebel testing. But, there are two pre-requisites. One is that STA should be enabled on the Siebel server and second is Siebel Add-in should be enabled in QTP.

Face to Face Interview–Round 2

Alex: I have loaded Siebel Add-in only and Web Add-in is not loaded. Can I still identify the Siebel application browser object?

Me: Yes, the Web and ActiveX Add-ins get automatically loaded with the Siebel Add-in.

Alex: What is Siebel Test Express?

Me: SiebelTest Express is an option in Object Repository Manager which allows connecting to the Siebel database and pull up all the objects into the OR in one go. This way we don't need to go the application for adding objects. But, using this approach may lead to unused objects in the OR which never get used by any of the tests.

Alex: What challenges have you faced while working with Siebel Applications?

Me: The most problematic thing that I faced was intermittent errors when the object is not on the screen. Once, it took me over 30 odd runs of the same script to figure out that this issue was mainly due to object not being visible on the current view of the page. Finally by increasing the resolution we were able to resolve it.

Another issue was searching sub-children in the List applet as we had to expand every node and then search for the value. The same is the issue when searching for a value. We were able to do it with a loop but the performance was poor and there was no option to fix it also.

One more issue was working with child items of the list applet as these don't get added to the OR directly and we had to manually define these objects. Moreover, because of the CAS API, we cannot use Descriptive Programming when working with the List applet.

Another issue was with modal dialog boxes where clicking on a button doesn't return the control back to the script and causes a timeout error. For this we had found two resolutions, one was to use the object as Web Test Object instead of a Siebel object and other was to click on the object using DeviceReplay.

These were the key issues specific to Siebel that we had faced.

Alex: Do you prefer using DataTable or Excel sheet for Data Driven Tests? And why?

Me: My preference is always to keep things simple as much as possible. So, if my requirements get satisfied by using just DataTable then I don't use Excel spread sheets in that case.

If my test requires sharing information across other tests at the same time, then I would go for an Excel spreadsheet. Also, in the case of requirements where colouring, formatting etc are required, I would prefer Excel.

Alex: We have an application in an Agile environment. Every day something is added or removed from the same. Object changes and flow changes are very common. How would you automate such an application?

Me: Well, I think the application breaks all the rules of being favourable for Automation. The requirements are not stable, there is high maintenance required. This would mean that every day the scripts or the objects may need to be updated. This is simply an overhead as the automation team will spend more time updating their scripts and GUI objects than running scripts to find defects.

Alex: Let's say ignoring all this I still want to automate this application then what will you do?

Me: I think such an application needs a framework that works out to be easily maintainable. In this case, I would move data into an Excel file or database. All object definitions, script flow definitions etc should be stored in Excel. The QTP script will be completely driven by the external Excel files which can be easily updated using keywords. Each keyword would map to a Function in the global library file. As and when flow changes, new Functions can be created and added to the framework and they can be called in the script's Excel files. But I guess having developers also included into this could make things easier.

Alex: You have executed Tests from QC. The application being tested is a web application. How do you make sure that the QC browser doesn't interfere with your script?

Face to Face Interview–Round 2

Me: From QTP 9.x there is a setting in Tools->Options to ignore the QC browser. By default, this option is checked and ignores QC. This way we don't need to worry about the additional QC browser.

Alex: Consider I run the test from QC which has below code

```
For i = 0 to 1000000
   Reporter.ReportEvent micPass, "i=" & i, "i=" & i
Next
```

Now the user hits the stop button when the script is running. What would be the status of the script in QC? Will it be "Failed", "Passed" or "Not Completed"?

Me: In case we stop a test in between then QTP will update the status of the test best on current state of script. If the test has already failed before then the status will be reported as "Failed". In this case since the test has only "Passed" statements, the test will show as "Passed" in QC.

Alex: Isn't this an issue, where even an uncompleted test script is reported as passed?

Me: Yes it is, but that is how it works. Though as a user, I would like if HP changed this and displayed the test as "Not Completed".

Alex: Can you think of a solution for this? The test should be failed when a user stops it.

Me: Whether the test is manually stopped or is completed successfully, all the global class objects created are destroyed. This means, the Class_Terminate destructor is fired for each of the class that executes during the session. So, we can create a class to check if user stopped the test.

```
Class CheckUserAbort
   Sub Class_Terminate()
     If IsTestAborted() Then
```

```
    Reporter.ReporterEvent micFail, _
            "The test has been stopped by the user", _
            "The test has been stopped by the user"
    End If
  End Sub
End Class

Dim CheckAbort
CheckAbort = New CheckUserAbort
```

Now, we need to implement the IsTestAborted method. To do so, we can use the below code in an associated library:

```
Dim IsAborted
IsAborted = True


Function IsTestAborted()
   IsTestAborted = IsAborted
End Function
```

Now we can update the code of test as:

```
For i = 0 to 1000000
   Reporter.ReportEvent micPass, "i=" & i, "i=" & i
Next
IsAborted = False
```

Face to Face Interview–Round 2

This way when a user aborts the test in between, the IsAborted flag will remain set as True and our Class_Terminate event will capture that and fail the test.

Alex: But your approach would fail when you have multiple iterations or multiple exit points in the test?

Me: *(Ahh gosh...I forgot I am up against a perfectionist.)*

To counter iterations problem, before setting the flag we check if it is the last iteration or not.

```
'Check if current iteration is also the last iteration
If  Environment("TestIteration")  =  Setting("LastGlobalIteration") Then
   IsAborted = False
End if
```

Alex: Can you think of any situation where this approach won't work?

Me: *(It's hard to find a bug this way...let me just think harder)*

Well this could fail if we import the DataTable at run-time as at that time the Setting("LastGlobalIteration") will not updated be with the new value.

Alex: So anything to fix that?

Me: I would need to try some code in QTP. And I would need 10-15 minutes for this.

Alex: Fine, you can use the laptop in front of you

Me: *(I remember the undocumented technique of enumerating Setting variable that I had read in "QuickTest Professional Unplugged". I thought that was my only chance to fix this up. I remembered the code in bits and pieces but it took 10 minutes to finally get the Function right)*

And I thought I knew QTP!

```
'ObjSetting is of Type
Public Function EnumerateSettings(objSetting)
    'Get arrays of key
    vKeys = objSetting.Keys

    'Get the 2nd dimenstion of the Keys array
    i_Count = UBound(vKeys,2)
    On Error Resume Next


    'Loop through all the keys and get the details
    For i = 0 to i_Count
        DataTable("Key",dtGlobalSheet) = vKeys(0,i)
        DataTable("Value",dtGlobalSheet) = objSetting.Item(vKeys(0,i))
        DataTable("Type",dtGlobalSheet) = TypeName(objSetting.Item(vKeys(0,i)))
        DataTable.GlobalSheet.SetCurrentRow i+2
    Next
End Function
```

I changed my earlier class code as below

```
Class CheckUserAbort
    Sub Class_Terminate()
    Call EnumerateSettings(Setting)
```

```
    End Sub
End Class
```

```
Dim CheckAbort
CheckAbort = New CheckUserAbort
```

I ran the test twice; once letting it complete and once aborting. The idea behind this was to see if any undocumented setting variable can be used to check for abort scenario. I saved the run-time DataTable from the test results and compared to see if there is anything that changes when we abort a test. I found something interesting named "onabort" and the value as False in one result and as True in the other. Jackpot!!!)

I think I have found a solution. QTP has one undocumented variable Setting("OnAbort") which it sets to True if the user aborts the test. The updated code would look like

```
Class CheckUserAbort
  Sub Class_Terminat()
    If Setting("OnAbort") Then
      Reporter.ReporterEvent micFail, _
        "The test has been stopped by the user", _
        "The test has been stopped by the user"
    End If
  End Sub
End Class
Dim CheckAbort

CheckAbort = new CheckUserAbort
```

(I knew I had found something that has never been used before J. Even Alex seemed a bit in disbelief that a solution of this kind exists. I was glad to have read my copy of QTP unplugged very thoroughly, otherwise it would have been impossible to find the solution)

Alex: How do you identify the color of a WebElement in QTP?

Me: There are no Test Object properties for this purpose supported by QTP, so we need to use DOM instead. We can use the currentStyle object which provides access to the stylesheet values and access the color property of the same:

```
.WebElement().object.currentStyle.color
```

Alex: There is style object also supported, why did you use currentStyle over here? Is there any difference between these?

Me: Yes there is a reason behind using currentStyle instead of style. Consider the below HTML:

```
<div style="display: none;">
<span style="color: red;">Test</span>
</div>
```

The Style object on the SPAN element will reveal only properties that were specified for the element. They will not indicate styles that might have been inherited from other parent objects. But currentStyle gives information on all the cascaded styles for the element.

Alex: What will be the output of below code?

```
Set oDic = CreateObject("Scripting.Dictionary")
oDic.Add "IN", "India"
oDic.Add "DEL", "Delhi"
oDic.Add "PUN", "Pune"
```

```
allKeys = oDic.Keys

Dim arrVals

For i = Lbound(allKeys) to UBound(allKeys)
   ReDim arrVals(i)
   arrVals(i) = oDic(allKeys(i))
Next

Msgbox Join (arrVals, ",")
```

Me: Let me just go through it

The output would be "Pune". The reason being we resize the array every time inside the loop but we don't preserve the old values using `ReDim Preserve`. So, on every resize, the values are emptied and the array has blank elements. But when the last iteration of the loop is executed the resize doesn't happen and the last key of dictionary remains intact in the last element of the array.

Alex: How do you get the title of the topmost window?

Me: The topmost window has the property foreground set as True. So we can use the same to recognize the window back:

```
Msgbox Window("foreground:=True").GetROProperty("title")
```

Alex: You have an application which uses the 'Always on Top' feature. It means at any instance of time the application doesn't allow any other window on top of it. You have created a QTP script for the same. How would you debug it?

Me: If a window remains always on top then we can resize the window to a smaller size and then debug the test through QTP. If that is not possible in the case where it is not allowed to resize the window, we can minimize the application and then debug the script. Also, another option could be to use dual monitors.

Alex: Does QTP support dual monitors?

Me: QTP 11 officially supports dual monitors but in lower versions we need to keep the AUT on the primary monitor and we can keep QTP on any of the monitors.

Alex: How do you close the last browser that was opened?

Me: The count of the browsers can be taken using ChildObjects on the desktop object

```
Set oBrwDesc = Description.Create
oBrwDesc("micclass").value = "Browser"
Set oBrowsers = Desktop.ChildObjects(oBrwDesc)
```

Now to close the browser we have two ways

```
oBrowsers(oBrowsers.Count - 1).Close
```

or we can use the creationtime property

```
Browser("creationtime:=" & oBrowsers.Count - 1).Close
```

Alex: There is a WebTable in an application having more than 10,000 records. This table is sorted. Now if you want to search for particular value within this table then what would be the best approach in terms of performance considering number of records exists within this WebTable?

Me: Since the list is sorted and we need better performance as well, I guess we should go for a Binary search where we start from the midpoint of the list. In case the value to be searched is greater than the value at mid point then we move higher in the list else we move to the left of the list. The code would be as below

```
iStart = 1
iEnd = ...WebTable().RowCount
Do While iStart < iEnd
    iMiddle = (iStart + iEnd) \ 2
    value = WebTable.GetCellData(1, iMiddle)
    iCompare = StrComp(value, strFind)
    If iCompare = 0 Then
        'Exact match has been found
        Msgbox "Value found at - " & iMiddle
        Exit Do
    ElseIf iCompare = 1 then
        'Value is greater than strFind
        iEnd = iMiddle - 1
    Else
        'Value is less than strFind
        iStart = iMiddle + 1
    End If
Loop
```

Alex: I have a WebTable in my application and I want to print all the rows in column 4 that have value "INDIA". How would you do it?

Me: The WebTable object provides a method called GetRowWithCellText which returns the row number containing a specified text. The syntax of GetRowWithCellText is:

```
WebTable("").GetRowWithCellText(Text, Column, StartFromRow)
```

We can use the same in this case. The code for the same would be

```
Set oTable = ...WebTable("Table")
iFound = 1
While iFound <> -1
    iFound = oTable.GetRowWithCellText("INDIA", 4, iFound)
    If iFound <> -1 Then Print "Found text at row - " & iFound
Wend
```

Alex: You said you have mostly worked with Web Automation, so you must be aware of HTML DOM functionalities as well?

Me: Yes, I am.

Alex: Ok, tell me the methods we can you use to get an object by its name?

Me: There are two way of getting a object by its name

```
Browser().object.document.GetElementsByTagName("Name")
Browser().object.document.all("Name")
```

Alex: Ok, tell me the methods we can you use to get an object by its id?

Me: Again, there are two way of getting a object by its id

```
Browser().object.document.GetElementById("id")
Browser().object.document.all("id")
```

Alex: What is the difference between GetElementById and GetElementsByTagName?

Me: GetElementById returns the object referred by the ID while GetElementsByTagName returns the collection of object matching the name, even if there is only one.

Alex: What will happen when there are multiple objects with same id, what would GetElementById return?

Me: It would return the first object with the matching id.

Alex: You mentioned the 'all' collection earlier, what would happen with that approach in case of multiple names?

Me: All will return the collection of object matching the name or id in such case. So if there are two objects with the name as "test" and 1 object with id as "test" then all would return 3 objects.

Alex: How would you work on the collection object returned by GetElementsByName?

Me: There are two options. We can either use a For loop or a For Each loop. Assuming oCol is the collection returned

```
For i = 0 to oCol.length - 1
    sHTML = oCol(i).outerHtml
Next
```

Or

```
For Each oObj in oCol
    sHTML = oObj.outerHTML
Next
```

Alex: How do you fire an event on the given object?

And I thought I knew QTP!

Me: Once we have the object reference, we can directly call the event on the object or use FireEvent to fire the event. Assuming oObj has the object reference we can use

```
oObj.ondblclick
```

or

```
oObj.FireEvent "ondblclick"
```

Alex: Assume you have an object reference to a given HTML node. Now somewhere in the HTML tree, it has TABLE node. How can I get the reference to this TABLE node?

Me: Each HTML node in DOM supports a property named tagName. Therefore, we can use tagName to check the name of the tag. Assuming oNode as the actual object reference we can use the parentNode property of the object

```
Dim oParent
Set oParent = oNode
Do
   Set oParent = oParent.parentNode
Loop While oParent.tagName <> "TABLE"
```

So at the end of the loop the oParent will contain the TABLE object node.

Alex: How would you find an Image with alt property as "continue"?

Me: Image object is presented using IMG tag. So we can use the below code to loop through all such objects

```
Set allIMG = Document.GetElementsByTagName("IMG")
Dim oIMGFound

For i = 0 to allIMG.length - 1
```

```
    If allIMG(i).alt = "Continue" then
        Set oIMGFound = allIMG(i)
        Exit For
    End if
Next
```

Alex: How do you get specified cell content from the TABLE node object?

Me: A table object has a row collection which represents the rows in the table and each row has a cell collection. To access cell content, we can use these two collections

```
cellContent = oTable.Rows(iRow - 1).Cells(iCol - 1).outerText
```

Alex: We spoke about the GetRowWithCellText method earlier, can you create a similar method which instead of the cell's text, retrieves the text of the entire row. Also, it should support regular expressions we well.

Me: We can do that using DOM only. First we need a Function which can test for regular expression

```
Function IsRegEqual(Text, Pattern, IgnoreCase)
    Dim oReg
    Set oReg = New RegExp
    oReg.Pattern = Pattern
    oReg.Global = True
    oReg.IgnoreCase = IgnoreCase

    IsRegEqual = oReg.Test(Text)
End Function
```

And I thought I knew QTP!

Now we can create a Function to iterate through each row and check the text if it matches the pattern

```
Function GetRowWithRowText(oTable, Text, StartFromRow)
   Dim oTableDom

   'Get the DOM Object from the WebTable object
   Set oTableDom = oTable.Object

   'Use -1 to search in all rows
   If (StartFromRow = -1) Then
     StartFromRow = 0
   Else
      'QTP indexes are 1 based and DOM are 0 based
      StartFromRow = StartFromRow - 1
   End If

   Dim iRow, sRowText

   For iRow = StartFromRow to oTableDom.Rows.length - 1
      'Get the text of the row
      sRowText = oTableDom.Rows(iRow).outerText
```

```
    If IsRegEqual(sRowText, Text, True) Then

      'We have found the row, return it

      GetRowWithRowText = iRow + 1

      Exit Function

    End If

  Next


  'No row found

  GetRowWithRowText = -1

End Function


RegisterUserFunc "WebTable", "GetRowWithRowText", "GetRowWithRow-
Text"
```

Alex: Can you create a custom method for WebList as well which can select values based on Pattern?

Me: The WebList DOM object supports an Options collection which allows accessing all the elements of the WebList

```
Function SelectUsingPattern(oWebList, Text)

  Dim oListDom

  'Get the DOM object of WebList

  Set oListDom = oWebList.Object

  Dim i, sOptionText
```

```
  For i = 0 to oListDom.options.length - 1
    sOptionText = oListDom.Options(i).Text

    'Check if value matches the pattern
    If IsRegEqual(sOptionText, Text, True) Then
      oListDom.Options(i).Selected = True
      Exit Function
    End if
  Next
End Function

RegisterUserFunc "WebList", "SelectUsingPattern", "SelectUsingPattern"
```

Alex: How can you some JavaScript on the current open web page?

Me: The window object in DOM has an execScript method which can be used to execute a JavaScript

```
oDocument.parentWindow.execScript "function sum(x,y) {return x+y; }"
```

Alex: Well, you declared a Function here but how would you call it?

Me: The variables or Function created using execScript can be accessed using the parentWindow object itself. We can call the Sum Function we created just now as below

```
MsgBox oDocument.parentWindow.Sum(2,3)
```

Alex: How do you count the number of links on a web page using DOM?

Me: There are two ways one is to use the links collection and other is to use GetElementsByTagName

Face to Face Interview–Round 2

```
Cnt = Browser("").Page("").object.links.length
```

Or

```
Cnt = Browser("").Page("").object.getElementsByTagName("A").length
```

Alex: Have you seen the Google auto suggest list during search?

Me: Yes. I have

Alex: Have you ever automated the selection from this suggestion list?

Me: No, I have only used during my searches but never looked into its Automation.

Alex: Have you done file handling in QTP?

(I wondered how did I get spared this time? I thought the next question that was coming was how you would Automate the search box)

Me: Yes. We can use 'Scripting.FileSystemObject' to work with files.

Alex: How would you copy content of one file to another file?

Me: We can use the CopyFile method for this

```
Set FSO = CreateObject("Scripting.FileSystemObject")
FSO.CopyFile "C:\File.txt", "C:\FileCopy.txt", True
```

Alex: This is the default method, I want you read the content from file first and then write it to a new file

Me: We can do that also

```
Set FSO = CreateObject("Scripting.FileSystemObject")
```

```
'Read the file content
Set oFile = FSO.OpenTextFile("C:\File.txt")
Dim sFileContent
sFileContent = oFile.ReadAll
oFile.Close

'write the file content to a new file
Set oFile = FSO.CreateTextFile("C:\FileCopy.txt")
oFile.write sFileContent
oFile.Close
```

Alex: How do you a read a file line by line into an array?

Me: We can use the ReadLine method for this

```
'Read the file content
Set oFile = FSO.OpenTextFile("C:\File.txt")
ReDim sFileLine(-1)
While Not oFile.EOF
   'Increase the array size by 1 and preserve all previous lines
   ReDim Preserve sFileLine(UBound(sFileLine) + 1)
   sFileLine(UBound(sFileLine)) = oFile.ReadLine
Wend
```

Alex: Your code does too many ReDim Preserve, how can you fix that up in the code?

Me: We can read the whole file in one go and then split it into lines

```
Set oFile = FSO.OpenTextFile("C:\File.txt")
Dim sFileLine
sFileLine = Split(oFile.ReadAll, vbNewLine)
oFile.Close
```

Alex: How do you print name of all the subfolders in a folder?

Me: We can first get the Folder object using the GetFolder method and then use the SubFolders collection to enumerate through each sub folder

```
Set FSO = CreateObject("Scripting.FileSystemObject")

Set oFolder = FSO.GetFolder("C:\Windows")

For each oSubFolder in oFolder.SubFolders
    Print oSubFolder.Name
Next
```

Alex: How about printing the file names?

Me: Instead of SubFolders collection we can use the Files collection

```
Set FSO = CreateObject("Scripting.FileSystemObject")

Set oFolder = FSO.GetFolder("C:\Windows")

For each oFile in oFolder.Files
    Print oFile.Name
Next
```

Alex: How do you find if a file contains certain text or not? And how do you replace the same with new content?

Me: For checking if the content exists or not, we can read the content and then use the InStr method. For replacing it, we can use the Replace Function.

```
'Read the file content
Set oFile = FSO.OpenTextFile("C:\File.txt")
Dim sFileContent, bFound
sFileContent = oFile.ReadAll
oFile.Close

sSearchText = "test"
sSearchReplace = "Best"

'if non-zero then the text is Found
bFound = InStr(sFileContent, sSearchText) <> 0
sFileContent = Replace(sFileContent, sSearchText, sSearchReplace)

'Write the replaced content back to file
Set oFile = FSO.CreateTextFile("C:\File.txt")
oFile.write sFileContent
oFile.Close
```

Alex: How can you find a file named MyFile.txt with in any folder of C:?

Me: We can create a Function to search file and then recursively in sub folders

```
Function SearchFolder(objFolder, FileName, Recursive)
   Dim oFile, oSubFolder

   For each oFile in objFolder.Files
   If oFile.Name = FileName Then
      'Found the file print the name
      Print "File found at - " & oFile.Path

      'No more files with same name can exist
      Exit For
   End if
   Next

   If Recursive Then
   For each oSubFolder in objFolder.SubFolders
      'Check in sub folders as well
      Call SearchFolder(oSubFolder, FileName, Recursive)
   Next
   End if
End Function
Set FSO = CreateObject("Scripting.FileSystemObject")
'Get the root folder C:\
Set oRoot = FSO.GetDrive("C:").RootFolder

Call SearchFolder(oRoot, "File.txt", True)
```

Alex: This would be time consuming if the folder depth is huge. Can you implement a solution which doesn't require you to loop through each folder and is better in performance?

Me: (I was stumped on hearing this as I couldn't imagine searching a file in less time with some other code)

To do a faster search we may need to build a custom database of all files and fire select queries on that.

Alex: Wouldn't you face the issue of keeping it in sync with the file system?

Me: Yes.

(I didn't know what he was expecting from me, I had no clue)

I am not sure how to do that. I know that Windows provides certain APIs like FindFirstFile, FindNextFile. But I have never worked with these APIs and also they use structures, so for sure they can't be used in QTP

Note: Nurat had the right thought of querying the file name through a database but creating such data source is not required. WMI, also known as Windows Management Instrumentation provides a way of firing SQL queries on various WMI classes and returns the object as a collection. To search a file in C: without using a loop we can use the below code,

```
Set objWMIService = GetObject("winmgmts:\\.\root\cimv2")


Set colFiles = objWMIService.ExecQuery _
    ("Select * from CIM_DataFile Where FileName='File' and Extension='txt' and Drive='c:'")
```

```
    For Each objFile in colFiles
        Print objFile.Drive & objFile.Path
        Print objFile.FileName & "." & objFile.Extension
    Next
```

Alex: I have an application which allows launching multiple instances. Now I want to launch two instances of this application and test them through QTP for comparison. How can I do this?

Me: We can just launch both the application instances using SystemUtil.Run and then identify one with index:=0 and other one with index:=1.

Alex: Let me make things a bit tough for you with this one. I don't want you to use index while identifying them?

Me: Will they have same titles or different titles in multiple instances?

Alex: They will have same title.

Me: (Same title. So he doesn't want me to identify the object using title and index. What can be unique then?? I guess HWND needs to be used now)

Even if the titles are same, both the windows will have a unique Windows Handle (HWND). We can use that to recognize the windows.

Alex: But how would you get the HWND of both the windows when you are not allowed to use the index?

Me: I was thinking of using Window("title:=App","index:=0").GetROProperty("hwnd") to get the handle of first and similarly next window but you restricted that out also.

(Without index....without index.....we need ChildObjects now)

And I thought I knew QTP!

QTP has a utility object named Desktop which supports a ChildObjects method. We can create a description having the title string and then getting the matching windows using 'Desktop.ChildObjects'. Afterwards, we can loop through the collection and get the handle of both windows without using INDEX!! In other words, I would do this to retrieve the Windows Handle for both windows:

```
Set oDesc = Description.Create
oDesc("title").Value = "App"


Set oParent = Desktop.ChildObjects(oDesc)


For ix = 0 to oParent.Count - 1
  Print "Handle of Window #1: " & oParent(ix).GetROProperty("hwnd")
Next
```

Alex: I have this code

```
Set oDesc = Description.Create
oDesc("title").Value = "Test"
Window(oDesc).SetTOProperty "title","Best"
Msgbox Window(oDesc).Exist(0)
```

I have the window with title Test and not with Title Best. What would be the output of above code?

Me: The Exist would return True. SetTOProperty has no use when using Descriptive Programming. Though the statement would run without errors, but it will actually not do anything.

Alex: Consider a WebList in the application has 10 items. I want to read all the 10 items in one go. How can I do that?

Me: We can use GetROProperty("all items") which would return all 10 items separated by a colon (;)

Alex: Which Test Object property determines if a Web control is disabled?

Me: It is the disabled property only.

Alex: How do you right click on a Link in a Web page?

Me: If we are using QTP 11 then we can use the RightClick method. Else, we can use the click method and pass the mouse button parameter as micRightBtn.

Alex: I have selected only Web Add-in during QTP start up. Now I want to automate on a windows based application. What will I need to do to make it work?

Me: Nothing. There is no Add-in needed for identifying standard windows based applications. Whether web Add-in is there or not, it will not make any difference. However, if the aim is to only automate on a standard windows application, there should be no need to select the Web Add-in.

Alex: How do you click a cell in a WebTable?

Me: In case there is no child table in the table then we can add on to the hierarchy using something like below

```
WebTable().WebElement("html   tag:=TR",   "index:="    &    (row-1)).
WebElement("html tag:=TD", "index:=" & (col-1)).Click
```

The trick here is to reach to the TD tag which represents the actual cell.

Another way is to use the DOM object on the table:

```
WebTable().object.Rows(row-1).Cells(Col-1).Click
```

Lalwani

Alex: How do you double click a Row in a WebTable?

Me: We can first get the WebElement of the cell using the approach I just told you and then we can use FireEvent method to fire the ondblclick event on the element.

Alex: I have a breakpoint set in QTP, when I run the Test, QTP still doesn't stop on the breakpoint. What could be the issue?

Me: For debugging the pre-requisite is that Microsoft Script Debugger should be installed. If it's not installed, breakpoints won't work. Also, if we have configured the Run mode of QTP to Fast, then again, breakpoints won't work.

Alex: I have a scenario where DataTable is imported in the start of the test and then exported back at the end of the test. Now I want to move my script in QC and have the DataTable updated in QC?

Me: DataTable.Import statement can take a local path as well as a QC path, but we can't use the same with DataTable.Export.

So to move such test to QC, we will have to add code that exports the DataTable to a temporary file, then delete the file from QC using OTA API and add the file again as attachment

Alex: Why delete the existing file?

Me: QC APIs don't allow overwriting an existing attachment. So we have to delete the file first.

Alex: Any issues you see with current approach?

Me: Yes, this approach won't work if multiple tests share the same DataTable and they are executed in parallel. This could create clashes where one of the DataTables may get overwritten. This can cause serious impacts on tests that are not yet executed and are in the cycle as the test would use invalid and/or incomplete data.

Face to Face Interview–Round 2

Alex: What is QC OTA API?

Me: QC OTA API means Quality Center Open Test Architecture Application Programming Interface. These API are installed with QC Client and expose COM interfaces for Automation. These API provide access to various database tables and objects from various tabs Requirement, Test Plan, Test Lab, Defects etc.

Alex: How would you write the code for downloading an attachment from QC? You can refer to the QC reference on the laptop if you want to

Me: Sure, give me two minutes. Here is the Function

```
'Function to Add an attachment to a specified object
Public Function AddAttachment(ByVal TOObject, ByVal FileName)
    Dim oNewAttachment

    'Upload the new file
    Set oNewAttachment = TOObject.Attachments.AddItem(Null)
    oNewAttachment.FileName = FileName
    oNewAttachment.Type = 1 'TDATT_FILE
    oNewAttachment.Post

    'Return the new attachment
    Set AddAttachment = oNewAttachment
End Function
```

To add the attachment to CurrentRun of the test we can just call it below way

```
Call AddAttachment(QCUtil.CurrentRun, "C:\File.txt")
```

And I thought I knew QTP!

Alex: This won't work if you have an existing attachment. How would you remove the existing attachment if any?

Me: We can create another Function that does it

```
'Function remove an existing attachment from an object
Public Function RemoveAttachement(ByVal FromObject, ByVal FileName)
   Dim oAttachments, oAttachmet
   Set oAttachments = FromObject.Attachments.NewList("")

   'No attachments removed
   RemoveAttachement = False

   For Each oAttachment In oAttachments
     If oAttachment.Name(1) = FileName Then
        FromObject.Attachments.RemoveItem (oAttachment)
        'Attach ment has been removed
        RemoveAttachement = True
        Exit Function
     End If
   Next
End Function
```

And we can add the same call in our AddAttachment Function.

Face to Face Interview–Round 2

Alex: Fine. How would you download an existing attachment?

Me: We can use the below code

```
'Function to download attachments from a QCObject to a specified folder
Public Function DownloadAttachment(ByVal FromObject, ByVal TOPath, ByVal fileName)
  Dim oAttachments, oAttachmet
  Set oAttachments = FromObject.Attachments.NewList("")


  'No attachments removed
  DownloadAttachment = False


  Dim FSO
  Set FSO = CreateObject("Scripting.Dictionary")


  For Each oAttachment In oAttachments
    If oAttachment.Name(1) = FileName Then
      'Load the attachment to local drive
      oAttachment.Load True, ""

      'Attachment was downloaded
      DownloadAttachment = True
```

Lalwani

```
        'Copy the file from temporary downloaded location to the TO-
Place folder
        FSO.CopyFile oAttachment.FileName, _
            TOPath & oAttachment.Name(1),True
        Exit Function
    End If
  Next
End Function
```

Alex: I want to store and run all my scripts from the local drive. There is a TestSet in QC which contains the Test with the same name, how can I run my scripts locally and update its status in QC?

Me: The first thing we need is this case the name of the TestSet where the dummy test cases are stored. To get the test set using name we can use below code

```
Function GetTestSetFromName(ByVal TestSetName)
   Set TDC = QCUtil.QCConnection


   'Get the TestSet factory
   Set TSFac = TDC.TestSetFactory


   'Put the filter for our test set
   Set oFilter = TSFac.Filter
   oFilter("CY_CYCLE") = """" & TestSetName & """"
```

```
    'Get the fitered test
    Set oTestSet = TSFac.NewList(oFilter.Text)


    If oTestSet.Count = 1 Then
       Set GetTestSetFromName = oTestSet(1)
    Else
       Msgbox "Error occured. TestSet not found or multiple test sets found"
    End If
End Function
```

Now we need a class which will capture the time of test and create a run when the test fails.

```
Class QCReportStatus
   Sub Class_Initialize()
      MercuryTimers("TestDuration").Start
   End Sub


   Sub Class_Terminate()
      MercuryTimers("TestDuration").Stop


      Dim sStatus
```

And I thought I knew QTP!

```
'Check the status of the Test
Select Case Reporter.RunStatus
   Case micFail
      sStatus = "Failed"
   Case micDone, micPass, micWarning
      sStatus = "Passed"
End Select

Set oTestSet = GetTestSetFromName("Unit Testing")

Set oFilter = oTestSet.TSTestFactory.Filter
oFilter.Filter("TS_NAME") = """" & Environment("TestName") & """"
Set oTestSetTest = oTestSet.TSTestFactory.NewList(oFilter.Text).item(1)

Set oNewRun = oTestSetTest.RunFactory.AddItem(Null)

'Lock the object
oNewRun.LockObject
oNewRun.AutoPost = True

'Get a unique name
oNewRun.Name = oTestSetTest.RunFactory.UniqueRunName


oNewRun.Field("RN_DURATION") = MercuryTimers("TestDuration")\1000
```

```
        'Set the run status
        oNewRun.Status = sStatus

        'We are done unlock the object
        oNewRun.UnLockObject
    End Sub
End Class

Dim oQCReporter
Set oQCReporter = New QCReportStatus
```

Alex: What are different types of parameters QTP provides for parameterization?

Me: There are Action Parameters, Test Parameters, Environment variables and the DataTable object.

Alex: What is the difference between Test Parameters and Action Parameters?

Me: One difference is in the way they are accessed. Test Parameters are accessed using the TestArgs object while Action Parameters are accessed using Parameter object.

Test arguments can be accessed by all Actions while Action Parameters can only be accessed by the specific Action.

Action Parameter can be passed from within the test while Test Parameters can only be passed externally.

Test Parameters are set in File->Settings->Parameter menu while Action Parameters are set using the Action properties dialog.

Alex: You said Test Parameters can be passed externally, how?

Me: When we click the Run button, there is a dialog shown for selecting the test result location. It also provides another tab to set the Input parameter values.

Other way is by using the QTP AOM. The Run method of the Test Object in QTP AOM takes an optional parameter for specifying these parameters.

Alex: I start a Test and Test execution has failed reporting the error that Function library is not associated with the Test, how do you associate it without stopping the Test execution?

Me: If the Test has reported this error then Test will fail in any condition and nothing can be changed. But if we still want to continue with the Test then only possibility of that would be loading these Libraries when the Test pauses or comes into debug mode. At that time in the command window, we can use LoadFunctionLibrary in case of QTP 11 and ExecuteFile in case of lower versions of QTP.

Alex: My Application launches on windows start-up. I want to Automate it using QTP what can I do?

Me: QTP always has a requirement that it should be launched before the application. But there is one exception to this for standard windows applications. They can be launched even previous to QTP and it would still work. In case the application is dependent on any Add-in like VB, ActiveX, Web etc, then it may not work correctly. In that case we will have to terminate and launch it again.

Alex: Why doesn't QTP work if the application is launched before?

Me: It should have been a design limitation I guess. From what I know of the working of QTP, each Add-in has a specific DLL hook which is injected into the AUT process and QTP then uses the same for record and replay of events. When QTP is launched these hooks are activated and all new process launched after launching QTP get hooked properly. Anything that is launched before QTP will not have these necessary hooks and hence QTP doesn't identify the application properly.

Alex: Then why does it work for a Windows based application?

Me: Windows itself provides a lot many API to work with standard windows application. These APIs don't need any hooks and can be called by any process. I believe QTP also uses these API calls to perform Actions on windows objects and hence it works even if the application is launched before QTP.

Alex: What is the difference between Step Into, Step Over and Step Out during debugging?

Me: Step Into jumps into the next line of code that needs to be executed and pauses on it. If it is a Function statement, execution pauses on the first statement to be executed inside the Function.

Step Over completely executes the immediate statement and then pauses on the next statement to be executed. In case current statement is a Function call then all the code inside the Function will get executed and execution will pause on to next statement after the Function call.

Step Out is used to execute all code until the previous call statement completes. So we are inside a Function and we Step Out then it would come out of the call statement that was used to call the Function and will pause in the next executable statement in the caller code.

Alex: How do you run part of a script?

Me: QTP provides functionality called 'Run from Step'. We can right click anywhere in the code, select run from step to start execution from that point. But a problem with this feature is that, QTP removes all the text before to the line and then executes the code. So if we have a Function before the line which is being called later the in code, then QTP will throw a Type mismatch error as the code pertaining to Function was deleted by QTP. This statement is especially true if Function Libraries are dynamically loaded from QTP tests. If the libraries are not loaded, part of the executed code may fail to perform as expected.

Alex: There is a Session ID that is assigned to every login session by the server. How do you handle such an application in QTP?

And I thought I knew QTP!

Me: When a user manually logs into the website they don't have to manually handle these Session IDs, they are automatically handled by the server and the browser. In QTP, we simulate just what a user does, so there is no need for handling any session related information.

Alex: On a web page I have a table which loads asynchronously, the page is loaded completely and still the rows of table are getting updated. How do you handle such situation where you don't even know the no. of rows?

Me: Handling asynchronous objects or events can be difficult and need a custom solution which may be dependent on implementation of the object in question. In the current situation I can think of row count stability as the test, where after every few seconds we take the row count and compare it with the latest snapshot. If the RowCount is same then we assume that table is loaded completely. We can tweak the polling time by running few tests and observing the optimum time which gives stability to the approach

```
oldRowCount = 0
newRowCount = GetCount from Table
While newRowCount <> oldRowCount
    oldRowCount = newRowCount
    Wait 3
    newRowCount = Get row count
Wend
```

The downside to this approach is that it can impact performance because of static wait-times and rechecking the count every few seconds. However, even though it impacts performance, it is one of the only workarounds I have been able to find with VBScript and the methods QTP offers.

Alex: What debugging techniques do you use in your script?

Me: Having enough information on failures during unit testing and final execution can impact the efforts in fixing an issue. There are few key techniques that I follow to improve debugging

- Proper logging of Functions, parameters etc. The first line of each Function logs the Function name and its parameter. All this is dumped onto a debug log file

- Logging of all actual and expected result in the report. Sometimes developers don't log the actual value on failure. I always try and make sure there is as much information as possible to be logged

- I extensively use breakpoints where I want to debug the code flow

- Instead of stopping test runs, I try and manipulate the values at run-time through command window or watch. This saves time to design the script and gives me insight to where bugs are being created at run-time with different sets of data.

These are the key things that I usually follow.

Alex: While trying to automate an Application, some of the objects in the application are being identified as WinObject. What can be done so that QTP identifies the object properly?

Me: When QTP identifies an object as WinObject, chances are that the object has been implemented in a non-standard way. As a first step, we can try to map the user-defined object class to an existing type of Test Object and see if it works fine. This can be found in Tools->Object identification settings.

If this doesn't help then the issue could be with the technology of the application. It may be the case that application is a .NET application and we have not loaded the required .NET Add-in. So, we should check for the technology of the application and see if that particular Add-in has been associated with QTP or not.

If nothing works then the only option we have is to use the WinObject method or use Virtual Objects.

And I thought I knew QTP!

Alex: How do you send an email if a test fails at the end of the test?

Me: There are two parts to this problem. One is checking the status of script when it ends. Second is to send the email.

To check the status at the end of script we can instantiate a class in global library which would get destroyed automatically when the test ends. This would fire the Class_Terminate event for the object. There we can check if Repoter.RunStatus is micFail, if it is then it would indicate that the test has failed.

Now to send an email, we can write custom code utilizing Microsoft CDONTS or Outlook. We can use the same to send the email.

Alex: How do you attach the result XML or the result folder together in the email?

Me: The problem with sending the XML file is that it can't be done from within the script. The reason being, if we put the code of sending email in the script and try to attach the result, it would throw an access denied error since the script is still running and the result XML is being used by QTP.

So it only leaves us with one option, which is to execute our script using QTP AOM and then once the script is finished, send the email with the attachment.

Alex: How do you add an image to Test Result summary?

Me: Reporter.ReportEvent method takes an optional parameter which takes the image path. We can use that optional parameter to add images to test results. But this method is only available in QTP 10 or higher.

Alex: How do you add HTML to Test results?

Me: There is no direct method for this. There are few undocumented methods though. Once is to use `&gt;` or `&lt;` symbols in the text and the HTML text after that:

```
Reporter.ReportEvent micPass, "Color Text", "&lt;<B>This is bold text"
```

> **Note:** The above approach doesn't work in QTP 11 anymore. But there is another undocumented method LogEvent for Reporter object which allows us to Report HTML text and also Add images to the Test Result Summary

```
Set oEventDesc = CreateObject("Scripting.Dictionary")
oEventDesc("ViewType") = "Sell.Explorer.2"
oEventDesc("Status") = micPass
oEventDesc("EnableFilter") = False
oEventDesc("NodeName") = "HTML Text"
oEventDesc("StepHtmlInfo") = "<TABLE border='1'>" & _
              "<TR><TD>Actual  Value</TD><TD>Tarun</TD></TR>" & _
              "<TR><TD>Expected  Value</TD><TD>Tarun  Lalwani</TD></TR>" & _
              "<TR><TD>Checkpoint         Status</TD><TD style='background-color:red'>Failed</TD></TR>" & _
              "</TABLE>"


newEventContext = Reporter.LogEvent ("Replay",oEventDesc,Reporter.GetContext)


Function AddFileTOReport(ByVal FileName, ByVal NodeName)
    Dim oDesc
    Set oDesc = CreateObject("Scripting.Dictionary")
```

```
    oDesc("ViewType") = "Sell.Explorer.2"
    oDesc("Status") = micInfo
    oDesc("StepInfo") = ""
    oDesc("NodeName") = NodeName
    oDesc("IsDirectEvent") = True
    oDesc("BottomFilePath") = Replace(FileName, Reporter.ReportPath & "\Report\","")
    oDesc("ShowTopFile") = True
    oDesc("EnableFilter") = False
    oDesc("Action") = "InActUser"
    oDesc("DllPAth") = Environment ("ProductDir") & "\bin\Context-Manager.dll"
    oDesc("DllIconIndex") = 206
    oDesc("DllIconSelIndex") = 206
    Reporter.LogEvent "User", oDesc, Reporter.GetContext ' Reporter.GetContext
End Function

Desktop.CaptureBitmap Reporter.ReportPath & "\Report\MyFile.png"
AddFileTOReport "MyFile.png", "My File"
```

The above methods should only be used in QTP 9.5 or lower. In QTP 10 or higher version the last parameter of Reporter.ReportEvent should be used.

Alex: How do we generate the result in HTML format?

Face to Face Interview–Round 2

Me: One way is to write Functions in our script which then send results in HTML format.

Another option is to activate the Log Media in registry. QTP reporters are all in form of Media, which is nothing but a DLL file. By default, QTP uses the Report media which generates XML. There is a similar media named Log which generates HTML results. By default, QTP keeps it disabled but it can be activated in registry.

> Note: Open Windows Registry (regedit.exe) browse to the below key
>
> HKEY_LOCAL_MACHINE\SOFTWARE\Mercury Interactive\QuickTest Professional\Logger\Media\Log
>
> Change the value of Active from 0 to 1

Alex: I have a Browser object in the OR which uses the CreationTime property for object identification. The current value of CreationTime is set to 0. There are two browsers open, one with CreationTime 0 and the other with CreationTime as 1 and title as "KnowledgeInbox". If I run the below code

```
Browser("Browser").highlight
Browser("Browser").SetTOProperty "title", "KnowledgeInbox"
Browser("Browser").highlight
```

What would happen?

Me: It would highlight the same browser both times. Though the code may seem fine but the SetTOProperty is the culprit here. When we use SetTOProperty on a property which doesn't exist in the OR then it actually has no effect.

So, to make this work we need to add the title property to the OR for the Browser object.

Alex: Consider the below code

```
Browser("Browser").SetTOProperty "title", "Knowledge.*"
```

And I thought I knew QTP!

```
Browser("Browser").Highlight
```

The 'Browser' object in the OR has a title property added as "KnowledgeInbox" and there exists a browser with the same title. What will happen when I run this code?

Me: It may or may not throw an error and that depends on the value setting in the OR. If the checkbox for regular expression value is set in the OR for the title property, then the above code would work. Else, the code won't work. There is no way to set or unset the regular expression property at run-time when using OR. We can only configure runtime descriptions with or without regular expressions with Descriptive Programming.

Alex: There is a webpage having multiple links which are identical. You have to click one of the links. How would you do it?

Me: Whenever there are multiple links, we just need the index of the link on which we want to click. Now since there are so many links there are has to be some corresponding text on the web page UI which helps a user select which one they want to click. So, we need to use a similar logic where using a reference object can give us the index of the target object. We can perform this Action using an approximation strategy with x and y coordinates or through the use of related anchors.

Another feature that we can use in this case is the Visual relation identifier where we can specify reference objects which would be located near this link. But this feature is only available in QTP 11.

Alex: There is a windows application that you have to Automate. The problem with the application is that its title always keeps on changing when working with it. Rest of the objects inside the application are easily identifiable but the window itself is a problem. How would you automate such application?

Me: There are usually three different approaches that I have observed to be working solutions in this situation.

One is to remove the title property from the identification property all together.

Regexpwndclass is also by default for window object and if it is unique enough from other windows then this can do the job. But the pre-requisite is that only one instance of application should be open in this case.

Second method works if we can recognize the first time title. So if the application is launched and the initial title is constant then it can be recognized. We can add the first window object to the OR and then get its window handle. This window handle remains constant for the application session. So we can now use this handle to identify the window.

```
Hwnd = Window("Static").GetROProperty("hwnd")
Window("Dynamic").SetTOProperty "hwnd", hwnd
```

Third option is that if we know what would be the title of the object at the current instance then we always use that title before taking any action on the object.

Alex: Can we load Libraries at run-time using QTP AOM?

Me: If we consider run-time as when our QTP script is running then No. By the time we execute the code global scope is already ready and hence QTP can't add anything to it. But if we are running the Test through an external VBS file then we can add the libraries just before the Test starts.

QTP 11 provides a new method named LoadFunctionLibrary which can be used for this.

Alex: I have an Excel sheet which has a small button on it, how can I click that?

Me: QTP's object identification fails on most of Microsoft tools like Word, Excel and Outlook. So we can't use normal objects in this case. One option could be to use virtual objects in case it is feasible. Another option is to see if there is a possibility to get access to object that we are interested in through Excel COM API. It may or may not be possible depending what Excel COM APIs can expose and what they can't.

Alex: Have you use relative paths in QTP?

Me: Yes

And I thought I knew QTP!

Alex: What are they used for and how do they work?

Me: Relative paths don't specify the complete path of a resource which could be OR, library file, recovery scenario file, or a DataTable. They specify a path relative to the reference path. The reference paths can be configured in Tools-Options-Folder Tab. By default the current test directory is always added to this. We can add an absolute folder path here or a relative folder path also. E.g. "..\Framework" would mean go to the parent folder of current script and have the Framework folder added from its parent to current path. We can have multiple folders added to this folder's options.

Once we have created these folder options we can also add files using relative path. When QTP needs these files it combines the relative path of the file with the path specified in the folder options tab and then checks if the file exists. As soon it gets the first match it uses the same.

This relative path is useful for creating script or framework which is easily portable from one machine to other. Well planned relative paths could mean no maintenance required even if the re-usable code and actions are moved to a different location.

Alex: How do you find the file at run-time in code using this?

Me: There is utility object named PathFinder provided by QTP for this. It takes in the relative path and locates the file and returns the first match

```
filePath = PathFinder.Locate("Data.xls")
filePath = PathFinder.Locate("..\DataSheets\Data.xls")
```

Alex: How do you write results to Excel?

Me: We can write everything in DataTable and then export the sheet using DataTable. Export.

But, in case we are looking at reports which use colors, fonts etc then, we need to create a few re-usable methods which can insert data in Excel using COM APIs.

Face to Face Interview–Round 2

Alex: Tell me different methods we can use to simulate keyboard events in QTP?

Me: One is to use the Type method supported by QTP's Test Objects.

We can also use SendKeys method of WScript.Shell object.

QTP also provides an undocumented object Mercury.DeviceReplay which can also be used. It has methods like SendString, PressKey, KeyUp and KeyDown which can be used for simulating keyboard actions.

There is also a windows APIs keybd_event which is for sending keyboard events. But this method is more complex in comparison to the other 3 methods I mentioned.

Note: Below code demonstrates how to use the keybd_event API to simulate key press events

```
'Public Declare Sub keybd_event Lib "user32" Alias "keybd_
event" (ByVal bVk As Byte, ByVal bScan As Byte, ByVal dw-
Flags As Long, ByVal dwExtraInfo As Long)

extern.Declare   micVoid,"keybd_event","user32"   ,"keybd_
event" ,micbyte,micbyte,miclong,miclong

'Private Declare Function MapVirtualKey Lib "user32" Alias
"MapVirtualKeyA" (ByVal wCode As Long, ByVal wMapType As
Long) As Long

extern.Declare micLong,"MapVirtualKey","user32","MapVirtual
KeyA",micLong, micLong

Const KEYEVENTF_EXTENDEDKEY = &H1
Const KEYEVENTF_KEYUP = &H2

Const KEYEVENTF_KEYDOWN = &H0

Sub KeyDown(KeyAscii)
```

```
  extern.keybd_event KeyAscii, extern.MapVirtualKey(KeyAscii,
0), KEYEVENTF_KEYDOWN, 0

End Sub

Sub KeyUp(KeyAscii)

  extern.keybd_event KeyAscii, extern.MapVirtualKey(KeyAscii,
0), KEYEVENTF_KEYUP, 0

End Sub

Sub KeyPress(KeyAscii)

  extern.keybd_event KeyAscii, extern.MapVirtualKey(KeyAscii,
0), KEYEVENTF_KEYDOWN, 0

  extern.keybd_event KeyAscii, extern.MapVirtualKey(KeyAscii,
0), KEYEVENTF_KEYUP, 0

End Sub

Const vbKeyControl=17

Const vbKeyEscape=27

Const vbKeyR=82

call KeyDown(vbKeyControl)

Call KeyDown(vbKeyEscape)

Call KeyUp(vbKeyEscape)

Call KeyUp(vbKeyControl)

Call KeyPress(vbKeyR)
```

Alex: What exactly ReplayType does and when do we need it?

Face to Face Interview–Round 2

Me: ReplayType is a setting for Web Add-in only. It specifies whether events on web object needs to be carried out through mouse and keyboard simulation or they need to be carried out the internal browser events. When operations are carried out through browser events few application may not replay as expected. E.g. If we type into a WebEdit and that enables a WebButton on the application, then with ReplayType set to browser events may not enable the button as necessary events are not fired when setting the value on to the WebEdit. In such a case we should use ReplayType as Mouse.

Alex: Why shouldn't then we always use ReplayType as Mouse?

Me: We shouldn't because of few disadvantages

- One is that mouse replay is slower than normal replay
- Mouse replay will not work if the machine is locked
- Mouse replay may not work reliably if the application loses focus in between events

Alex: Is there a similar setting for Windows objects as well?

Me: No. For Windows objects we can use the Type method instead.

> *Note:* The setting does exists for changing replay type for the WinEdit objects, but it is not documented in QTP help. Below line of code will activate keyboard replay for setting the text on a WinEdit

```
Setting("Packages")("StdPackage")("Settings")
("ReplaySetTextWithType") = 1
```

Alex: Talking about the locked machine, can we run our scripts on a locked machine?

Me: Partially, not fully. If we run a script on a locked machine then few issues that we might face are

- QTP may not be able to identify objects. Though it would still be able to identify all web objects

Lalwani

And I thought I knew QTP!

- Any screenshots capture using CaptureBitmap will give a black screen
- QTP will fail on performing actions on non-web object
- It will also fail on web objects if the ReplayType settings is configured to mouse
- QTP test result summary will also show a warning that test was run on a locked machine. But we can disable this from registry by setting a the value of SkipEnvironmentChecks to 1

Such issues mostly make it difficult or rather impossible to run a test on locked screen

Note: The value SkipEnvironmentChecks is located at below path in system registry

HKEY_LOCAL_MACHINE\SOFTWARE\Mercury Interactive\QuickTest Professional\MicTest

Note: To perform operations on Windows objects when the machine is locked, we can't use the usual methods. In such a case we can rely on using the Windows messaging architecture. We can find the handle of the object we want to work on and send the appropriate message to it.

Consider the below code which sends a click event to a Button

```
Const BM_CLICK = &HF5
hwnd=Window("Explorer").WinButton("OK").GetROProperty("hwnd")
'Send the click event to the button
IResult = extern.PostMessage(Hwnd, BM_CLICK, 0,0)
```

Windows Messages reference for various controls can be found on the link below:

msdn.microsoft.com/en-us/library/bb773169%28v=VS.85%29.aspx

Note: To unlock a machine before starting the script execution, we can use Logon utility. The utility is available on the link below:

www.softtreetech.com/24x7/archive/sl.htm

 Note: We can make use of Virtualization to create different virtual images and run the scripts on them. This way we can keep the main machine locked but all the virtual machines running on the same will run unlocked.

Alex: What are Virtual Objects?

Me: Virtual objects are used when QTP is not able to recognize the object at all. Then we can map the region of the object which would be specified by its co-ordinates, width and height. This map is done to a Virtual object, which can be chosen as checkbox, button, table, list, radio button.

Alex: How do you use them in your Test?

Me: There is a wizard for creating Virtual objects. Once we have created the virtual object and we record on the application QTP will learn the virtual object. Usually I don't record so I don't prefer this method. Rather I use Descriptive Programming to do this.

Alex: How do you it in DP then?

Me: Based on the object type, there are different mandatory properties. Like VirtualButton object has x, y, width and height as the mandatory properties. We need to create the description for this and then use the same

```
Window("Test").VirtualButton("x:=10", "y:=10", "width:=10","height:=10").Click
```

Alex: How do you combine results of different scripts into one?

Me: I am not aware of any method as of now. But QTP stores the results in XML format and it completely defines the XML format, so there might be a possibility to write a script to merge XML from two different scripts and create on results XML. But I would never go in for such an approach as future of this approach will have dependency on HP not changing the XMLs.

Alex: How many folders are created when you save a QTP test?

Me: A test will have a total of 3 + 2 times the number of Actions. There is a folder for the main test and a folder for the Test flow which is Actiono. All other Action folders belong to individual Actions within the test. There will also be additional folders named Snaphots in each Action which stores the Active Screen information if any.

Alex: What all other files does a Test store?

Me: Each Action folder has three main files. Script.mts stores the Action code, ObjectRepository.bdb stores the local Actions object and Resource.mtr, which stores the parameter definitions and other information related to the Action.

Lock.lck in the main folder specifies if the test is locked by any other user or not.

Test.tsp file is more of an indicator to QTP that the folder is a QTP script.

Parameters.mtr stores the Test arguments and other information.

Default.xls stores the DataTable of the test.

There are few other files as well that are used by LR but I don't remember their names.

Alex: How do you upgrade all your Test scripts from existing version to a new version of QTP?

Me: One way is to open the test manually and save it. In case we have too many scripts then we can use QTP AOM to just open the script and save it again. The AOM script would be very simple

Face to Face Interview–Round 2

```
Set QTP = CreateObject("QuickTest.Application")
QTP.Launch
QTP.Visible = True
QTP.Open "C:\TestUpgrade"
QTP.Test.Save
```

Alex: I have QTP on Machine A and my application on Machine B. From Machine A I connect to Machine B using Remote Desktop. My application resides on Machine B. How can I automate such application setup?

Me: We can't. Remoting softwares usually render the remote machine more as an image rather than actual rendered controls. So QTP cannot identify anything inside the Remote Desktop window. The thumb rule for QTP is to have QTP and the AUT on the same machine.

Alex: So how can we use Remote Desktop with QTP?

Me: The application and QTP can be installed on the remote machine. Once, both QTP and the AUT are installed, we can connect to the remote machine using Remote Desktop and initiate the run. One issue that needs to be taken in this is that, a remote session cannot be disconnected once the execution is on and it can't be minimized also. If Remote Desktop is minimized then QTP won't be able to capture any screenshot and it won't be able to identify the object also.

This however is not a limitation in QTP 11 where we can change some registry settings and still have the test run on a minimized Remote Desktop.

Alex: Can we open a Remote Desktop connection run a script and close the Remote Desktop window?

Me: No we can't do that as it would be as good as locking the machine. And QTP scripts don't work on locked machines.

And I thought I knew QTP!

Alex: What are the advantages of associating Libraries to a Test?

Me: Associating libraries allows us to debug Libraries if there is an error. QTP doesn't allow debugging library load at run-time using ExecuteFile. But QTP 11 has a new method named LoadFunctionLibrary which allows loading libraries at run-time and allows debugging the same. Also, associated Libraries are loaded first before any Action which can help us execute some initialization code at startup. Also, all Functions and public variables of the global Library are available across all Actions.

Alex: How do you Maximize a Browser?

Me: Every top level object on the Desktop can be represented using a Window object. Hence by taking the handle of the Browser, we can identify it as Window object and then use the Maximize method of the same:

```
Hwnd    = Browser("X").object.HWND

Window("hwnd:=" & hwnd).Maximize
```

 Note: In case of IE7 and higher the above code may not work as they have a different hierarchy because of tab support. To work with the same one should use the code given below:

```
Dim hwndBrw, hwndWindow

hwndBrw = Browser("Browser").GetROProperty("hwnd")

Const GA_ROOT = 2

'Declare Function GetAncestor Lib "user32.dll" (ByVal hwnd
As Long, ByVal gaFlags As Long) As Long

Extern.Declare    micLong,    "GetMainWindow",    "user32"
,"GetAncestor",micLong, micLong


'Get the main IE window handle
```

```
hwndWindow = Extern.GetMainWindow(hwndBrw, GA_ROOT)
Window("hwnd:=" & hwndWindow).Maximize
```

Alex: How do you test sort order of a column in a table?

Me: We can just compare two consecutive elements and see if they are greater or lesser, if the same relation continues the list would come out to be ascending or descending. Pseudo code would be as below

```
bAsc = True
bDesc = True
For i = LBound(arrLOV) to UBound(arrLOV) - 1
   'Check if the order is still ascending
   bAsc = bAsc And (arrLOV(i) < arrLOV(i+1))
   'Check if the order is still descending
   bDesc = bDesc And (arrLOV(i) > arrLOV(i+1))
   'If it is neither ascending nor descending we can exist
   If Not (bDesc or bAsc) Then 'The list is not sorted
Next
```

Alex: I have a QTP Test Object and I want to know it's name in the OR, how can I do it?

Me: QTP doesn't provide any method to access the logical name of the object, but every Test Object in QTP supports a ToString method. From observation we know that it returns the logical name of the object and then the type of the object. Though it is very risky to use this approach as QTP doesn't document that ToString is supposed to return this and this doesn't work across all Add-ins. So if tomorrow they change this behaviour then this approach won't work.

Alex: Can you give some code that would extract the name from whatever approach you told me just now?

Me:

```
Function GetTestObjectName(Obj)
    GetTestObjectName = ""
    If IsNull(Obj) or IsEmpty(obj) Then Exit Function
    Dim strObjectName
    strObjectName = obj.ToString
        strObjectName = Split(strObjectName, " ")(1) & "(""" & Split(strObjectName, " ")(0) & """)"
   GetTestObjectName = GetTestObjectName (Obj.GetTOProperty("parent")) & "." & strObjectName
End Function
```

What I have done here is the take object and I do a ToString on that and then I get the object Name as well as type. I split the string with space and use the 1st element as OR name and 2nd element as object type. The code is raw and assumes that there are no spaces in the logical name and would give an extra dot also at the start. But I have written this to just show you the approach these small things can easily be fixed.

Alex: How do you get current Tests folder's path?

Me; We can use the built-in Environment variable Environment("TestDir") which would give the current Test's directory.

Alex: How to do create PDF of QTP's test results?

Me: This can be done through Test Result viewer tool only. This is only available in QTP 10 or higher version.

Alex: What is Active Screen and how is it useful?

Me: Active Screen stores a design-time image of every object in the application and allows you to add CheckPoints etc from the active screen itself. I feel the Active Screen is generally underutilized mostly because I feel recording is not the best way to automate. These active screens are only created during recording or when doing an 'Update Run Mode'.

Alex: Have you worked on Flex Automation?

Me: No, I haven't yet worked on Flex Automation. But the only thing I know about the same is that there is some file that we need to incorporate in our development code for QTP to be able to recognize these objects. But I am not hundred percent sure of that.

I think there are two plug-ins that are required to be installed – one for QTP and the other for the Browser object. Also, there is support only for IE6 as of now and a generic Flex hierarchy is the following:

```
Browser("").FlexApplication("").FlexButton("").Click
```

There are also several limitations:

- *Flex object do not show up in Tools -> Object Identification*
- *Unlike other Add-ins, we cannot find Flex objects help in QTP Object Model Reference*

Alex: How do you open a link in a new window?

Me: When we press the shift key and then click the link it gets opened in a new window. So we can send the shift KeyDown event on the browser window. And then we click on the link and then send the shift KeyUp event

The other thing is that we also need to change the ReplayType to mouse so that the click event is simulated through mouse and not browser events.

Alex: How can I use a variable from one Action to another?

Me: There are different ways to share data from Action to another Action. The methods that I have at top of my mind are:

- *Use variables in global scope. Once a library is associated with a test, all code within the library is accessible to all Actions within the test. Therefore, a variable that is declared Public within the library can be accessed by all Actions. For example, Action1 can store the value 'John' in the variable 'userName' and login with that userID. The same variable can be used in Action2 to verify if the correct user logged in.*

- *Use Environment Variables.*

- *Use Input and Output Action Parameters.*

Alex: Msgbox is a VBScript Function. How would you override this Function in QTP?

Me: We can define the Msgbox Function again in code as below

```
Function Msgbox(text)
   Print "Msgbox - " & text
End Function
```

But there is one catch here; if we define this in the associated Library then new Function will be only visible in the global namespace. In any of the Actions call the MsgBox Function then they would indeed refer to the default MsgBox method. This is because Actions run in different namespaces. Everywhere where we need to override the Function we need to define the new method.

Alex: Can we do something to keep the Function in Library itself and not define it again and again?

Me: Let me think

(If we want Function to be in one place then we can use Function pointer...)

For this we can use Function pointers. We can change our previous code as

```
Function NewMsgbox(text)
   Print "Msgbox - " & text
End Function

Dim ptrMsgBox, MsgBox

'Get the reference to new function
Set ptrMsgBox = GetRef("NewMsgbox")

'Override the message box now
Set MsgBox = ptrMsgBox
```

Now in every Action we just need to add two lines at the top to override the Msgbox method to our new one

```
Dim MsgBox

'Override the message box now
Set MsgBox = ptrMsgBox
```

This way we can have the Function only in the library file and use its pointer everywhere to override the Function.

Alex: Well you created a variable ptrMsgBox here to store the Function pointer. Why didn't you just use GetRef("NewMsgBox") directly in the Action?

```
Dim MsgBox

'Override the message box now
Set MsgBox = GetRef("NewMsgBox")
```

And I thought I knew QTP!

Me: There was a specific reason for not using this. The GetRef method can only get reference to a method in current namespace. So if we use GetRef in the Action then it would expect the NewMsgBox Function to be present in the Action itself.

Alex: Is there a way you can override a Function or method of an existing object? Consider the Reporter object in QTP. I want to add another method to it say ReportHTML and I want all existing methods also to be there. How would you do this?

Me: We can use a similar approach and this time instead of GetRef we will use an actual class object. To do so first we will create a new class which has all Reporter object methods and also our new ReportHTML method

```
Dim oOrgReporter, oNewReporter

'The original reporter object
Set oOrgReporter = Reporter
Set oNewReporter = New NewReporter

Class NewReporter
   'ReportEvent method
   Function ReportEvent(Status, EventName, Description)
      ReportEvent = oOrgReporter.ReportEvent (Status, EventName, Description)
   End Function

   'Getting the current filter value
   Property Get Filter()
      Filter = oOrgReporter.Filter
```

```
    End Property

    'Setting a new value for the filter
    Property Let Filter(newValue)
        oOrgReporter.Filter = newValue
    End Property


    'Run Status is read-only property, so we define the Get property only
    Property Get RunStatus()
        RunStatus = oOrgReporter.RunStatus
    End Property


    'ReportPah is read-only property, so we define the Get property only
    Property Get ReportPath()
        ReportPath = oOrgReporter.ReportPath
    End Property


    Function ReportHTML(Status, EventName, HTMLText)
        'Code to report the HTML text
    End Function
End Class
```

Now when we associate the above code in an associated library we have the new reporter object in the oNewReporter object with our added method. To use the updated reporter object in our Actions we can add the below-mentioned two lines at the top of the Action.

```
Dim Reporter
Set Reporter = oNewReporter
```

This way any code in our Action that use methods or properties of the Reporter object will be re-routed to our class object. But there is one thing to be noticed here, if we call any Function in the associated library then that Function would still use the original reporter object and not the overridden object. This is because we need at least one namespace where we capture the Reporter object. If we define the Reporter object again in global namespace as well then we don't we be able to get anything in the oOrgReporter. The below code demonstrates the issue

```
Dim oOrgReporter, oNewReporter


'The original reporter object
Set oOrgReporter = Reporter


Set oNewReporter = New NewReporter
Dim Reporter
Set Reporter = oNewReporter
```

When the code execute the line 'Set oOrgReporter = Reporter' we would expect the original Reporter object to be stored in oOrgReporter. But VBScript processes all variable declaration in the start. Hence our 'Dim Reporter' would already have been defined with an Empty value. That is why this approach won't work in associated libraries

Alex: I am sure you can pull some workaround to fix that also?

Me: (Phew! I knew it was coming....I have to declare the variable Reporter and still not have it declared before using it)

Let me think for a minute

(I can't directly declare the variable for sure, I need to declare it at run-time somehow. Oh Yes! Got it!)

Yes, there is a workaround. We can change the code as below so that the local Reporter object gets declared at run-time

```
Dim oOrgReporter, oNewReporter

'The original reporter object
Set oOrgReporter = Reporter


Set oNewReporter = New NewReporter
Execute "Dim Reporter"
Set Reporter = oNewReporter
```

Alex: There are multiple versions of an application. How do you manage?

Me: There can be multiple environments where an application may be running, but in my experience, all environments have somewhat mirrored one another. Therefore, a good automation script would run in either of the environments without any issues after a little bit of testing in each.

I generally have both Input and Output Excel spreadsheets for different environments. This is because data that is fed to the application would be different in each environment and the data coming out would be different as well. The same business logic drives my tests in each of the environments, but the data comes from different sources.

Because the logic to control different sources can be easily included in a driver script, it becomes easier to keep a single copy of the test script but multiple copies of data sources.

In projects where I have had access to Quality Center, it becomes even easier as any user can select the environment they want the automation script to be run and execute them from the Test Lab without having to open QTP manually.

Alex: Have you used a pencil?

Me: (I just remembered that it's been eight long years since I used a pencil but what would a pencil has to do with QTP???)

Yes, of course..

Alex: What ten uses can you think of a pencil except writing?

Me: (I knew where this was going now. I took a deep breath, I knew I had to break away from the technical questions and be very innovative to answer this question. I just relaxed for few seconds before starting. I knew I couldn't be hundred percent logical now.)

- Well since I am from QA, I would see a pencil as input for testing a sharpener
- We can use pencils in decorations
- We can use pencils for drawing (not same as writing)
- We can use the pencil waste also for decoration
- We can also use it for scratching our back

(Pause for 5 seconds...)

- We can also use two pencils as chop sticks
- We can use it as a scale to draw straight lines
- We can use it to test erasers as well

Face to Face Interview–Round 2

- We can play pencil fight games

(Two more to go and I was running out of ideas now...I thought hard for another 30 seconds)

- We can use dark pencils for makeup or eye lashes (Don't try this at home :P)

These are all the use I can think of as of now..

Alex: Okay. If you were a fruit, what kind of fruit would you be? Why?

Me: *(I didn't even take a second to answer this)*

I would be a Mango, I love mangos and I know many others do. Nothing better than to be a favourite of many people.

Alex: Consider that the distance between earth and moon is 1 million km and I give you an infinite size of paper. The thickness of the paper is 0.01 mm. Every time we fold the paper into half the thickness doubles. How many folds would you need to cover the distance between earth and moon?

Me: *(I knew there was some limitation to folding paper x number of times, I didn't remember if it was seven folds or eight folds. I thought rather than answering the folds I would just try something different)*

Sir, at one place we want to save trees and this question would require us to cut all of them.

Alex: (With shocked eyes!!) Assume you have that much of paper available

Me: *(I knew I had to give the count of max folds)*

Can I have an A4 size paper?

Alex: Sure, here take this one

Me: *(As soon as he gave me the paper I started folding it)*

Lalwani

And I thought I knew QTP!

This took seven folds because the paper is thick. It may go few more folds in case the paper is very very thin. So in any case we can't cover the distance between earth and moon this way.

Alex: If you are given a chance to become one of the existing super heroes, which one would you choose? Why?

Me: Well I would choose Batman. Because considering all the laws of physics and humanity, he is the one who looks relatively practical and possible.

Alex: What would you do if I give you an elephant?

Me: I would bill you for all its expenses

(I didn't ask for the trouble why should I pay for it…)

Alex: If you won a lottery of 10 million dollars what is the first thing you would do?

Me: I would collect the prize money first.

Alex: If a taxi and a CMW is priced the same, which one would you buy? Why?

Face to Face Interview–Round 2

Me: If I am taxi driver by profession I will buy a taxi else I will buy a CMW.

Alex: (Smiling) Nurat, I must say you have a good sense of humour.

Me: Thank you!

Alex: I guess we have reached the end of the interview. Would you mind waiting outside for a while.

I was again outside the meeting room waiting for my next update. I could feel the pain in my back due to sitting in the same posture for more than four and a half hours. It was one hell of an interview and I was completely drained out. I started stretching myself to feel better. While waiting, I was recollecting what I had gone through since morning. They took my interview as if NASA had been recruiting someone to send to Mission to Mars. They wanted to know everything and leave nothing for later. It would have been better if they could have done a brain scan for QTP, at least I would have been spared the pain. Beside such peculiar thought so many other things were coming out and I thought of writing a book just from the questions asked in this interview and I told myself perhaps not a bad idea to actually do it. There is no tax on dreaming to be an author. I was hoping the next round would be the final HR interview but no one had actually told me what's next. Alex had only asked me to wait.

Alex then came out of the room and told me that I would have HR round anytime in next thirty minutes.

I was called in by a lady for the interview in the same meeting room.

HR Interview

৸৩

INTERVIEWER #4: Nurat, Have a seat please. I am Ekta.

Me: Hi Ekta.

Ekta: Why do you want to leave Sysfokat?

Me: I have been with them for over seven years now and I believe a change would only better my career as I will get a chance to look at things from a different perspective. I know about the processes we have in Sysfokat and by moving to a new place I will get an opportunity to learn new things and will also be able to apply my previous knowledge as well.

Ekta: And tell me why do you want to join us?

Me: It has always been a dream for me to be with the company which revolutionized the way computing is done today. Working here with experts from different technologies would be a privilege and would help me shape my career.

Ekta: Is money the actual reason for this change?

Me: It is never a primary reason for me. But just like anyone else, I too want to improve my remuneration from time to time.

Ekta: How much hike are you expecting?

Me: I am expecting forty percent and above.

And I thought I knew QTP!

Ekta: I can give you a twenty-five percent hike. Are you fine with that?

Me: *(I didn't know what to say, so I thought maybe it is better to say No in a positive manner)*

No, not really.

Ekta: But just now you said money is not your primary reason for change?

Me: Yes. That's true. I am not a person who would want to change his company every one or two years and earn salary hikes. I prefer to have a stable job and if I can't get a better hike I would rather prefer to stay with my current employer. I may lose the additional 25 percent hike you are offering but as I said money alone is not decisive factor for or me.☺

(As soon as I said that I realized though I was honest with her but it may look like I was being over smart. I didn't know if this answer would have any negative impact on my interview. But I knew I said what I had in my mind.)

Ekta: Hmmm, Do you have any question for me?

Me: *(I wanted to ask why my technical interviews were so protracted. But I decided to drop that question. I was exhausted and just wanted to go back to the hotel.)*

Nothing specific, just wanted to know if there would be any more rounds of interview.

Ekta: (She smiled!) No, just fill this form up for me and then you can go back to your hotel. We will get back to you in a few days.

(I didn't know if something went wrong here, "will get back to you" is usually a polite way of saying NO.)

Me: Okay.

I filled up the form and handed it back to her. I then left the room.

I walked back to my hotel. It felt like the longest walk I had taken in a long time. I reached the hotel at around 10:25 p.m. I was so tired that I didn't even bother to have my dinner and promptly fell asleep. I woke up early morning at 6:00 a.m. with my stomach growling. I looked at my eyePhone and there were 10 missed calls and 8 of them were from Mom. I had changed the profile to silent mode before the interview and with all the hectic schedule of interview I even forgot to call up Mom. Since it was really early I decide not to disturb her. I ordered my breakfast in the room. I left the hotel at 7 o'clock as I had to catch my flight at 9:30 a.m.

I reached the airport at 8:15 a.m. and there was a huge line at the check-in counter. Luckily they had removed the pre-baggage screening at the new terminal in Delhi or else I would have surely missed my flight. At around 8:50 a.m. I got my boarding pass and went for the security check-in. The security check also had a huge line and it was only by 9:20 a.m., that I could make it through. I could hear announcements for last boarding call and I started running towards gate #11.

The airline staff asked me to board the bus asap and I was alone in the bus. I entered the plane and everyone was already seated. I decided to call mom quickly to let her know that I was leaving.

Me: Hi Mom!

Mom: Where are you? I called you up so many times yesterday

Air hostess: Sir, can you please switch off your phone?

Me: Mom, my plane is taking off now. I will call you later tonight. Bye.

Post Interview

The next day was Friday, a day when the office takes on a more cheerful look as we are allowed to dress in casuals. I reached my desk and opened Outlook to check my emails. When I checked my inbox, it felt like I had been away for ages although I had been on leave for two days only. Two Hundred and Twenty Six unread emails and by the time I finished reading them all I realized two hundred of them were just forwards or mails that weren't of any interest to me. I thought to myself, 'what a waste of time' but then suddenly lightning struck me. I realized I was on bench and I would need such things to kill time.

It felt really strange to be on bench after seven years of working on tight delivery schedules. I won't say it felt great as I always preferred to work but I knew I won't mind this break for a few days.

It was time to catch up with my friends and on the lunch table they had only one question, "Where were you for the past two days?". I somehow managed to change the topic and avoid answering their queries. They knew I was up to something but they had no idea what.

I spent the rest of the day reading some articles that I had saved on my desktop. Some of the articles were so engrossing that hours passed by before I realized that it was almost time to go home.

On reaching home I decided to watch my favourite movie on DVD — 'The Matrix'. I just loved the concept of the movie. The most interesting question the movie raised was regarding our interpretation of reality.

And I thought I knew QTP!

I was already halfway through the movie but at my back of my mind all the events from the past week were being replayed. There was this one thing that I knew might change the whole game and I was afraid that I might have got it all wrong. But I knew there was nothing I could do now as it was a thing of the past. If it was just one person's decision then I knew it would have been in my favour but there were many people involved in this.

Suddenly my eyePhone started beeping again and got me back from my thoughts. It was a Delhi no. again and I was ready with some funnier teasers this time. I picked up the phone, and…

Me: Hello!

Caller: Am I talking to Mr. Nurat?

Me: Yes

Caller: Hi Nurat, I am Ekta calling from MecroHard HR. I have called up to congratulate you as you have been selected. I will be sending you the offer letter in a few minutes.

Me: (I wondered if I was dreaming and I didn't reply for a few seconds.)

Ekta: Nurat, you there?

Me: Yes Ekta, I am sorry it seems the signal is a bit poor here. Thanks a lot.

Ekta: I want you to send the signed copy of the offer letter asap.

Me: Ekta, but when am I expected to join?

Ekta: In one month.

Me: Okay

(My company had a notice period of three months and I didn't know if I would get relieved in a month or not. But I didn't want to negotiate with the HR for now. I was just happy to get this opportunity.)

Post Interview

I checked my inbox and there it was — the offer letter from the MecroHard HR. **I had made it**. I still wasn't convinced. I read and re-read the mail several times to make sure the idea of being accepted sank in. It finally did. Not only did I clear the interview but I got an offer slightly better than what I had asked for.

I had almost a month to prepare for my move. It was going to be a very short month as I had a lot to do before I changed cities. I was extremely excited to take a new step in my life and I wanted to make the most out of it. I hoped I would rise to the challenges at my new workplace, maybe do even better.

One Month Later

One Month Later

༨༠༠༣

*F*inally, I joined MecroHard after going through a tough month. I really had a hard time convincing my now ex-employer to release me within a month. But I did manage to get released within a month.

My family and I were excited about me being back in Delhi, my hometown. I was about to start the new year with a new company and was really excited about the same.

I joined the MecroHard office and the initial 2-3 days were typically full of induction sessions. The sessions bored me as I was waiting to get down to the core work and meet the team I will be working with. On the fourth day Alex came and introduced me to the team. It was a small team of six people including Alex. He had scheduled an official meeting for the introduction as well at 10 o'clock. We all gathered in the meeting room, the same one where I was interviewed. All the memories of the interview day came rushing back in my mind as if the incident had happened just yesterday.

Alex: Hello everyone, I want to introduce Nurat; he comes here with seven years of experience in his kitty and is very good with QTP.

(Suddenly all of them started laughing. I wasn't sure what was funny about it. But I also tried smiling.)

Prasad: How about a demo?

(And everybody in the room started cheering.... Demo!!!!.... Demo!!!!... Demo!!!!. I suddenly remembered the days of my college ragging; this was no different. But I was excited that I would be part of an enthusiastic team.)

And I thought I knew QTP!

I looked at Alex and was searching for an answer and...

Alex: Why not!

(Those words made me nervous, and I was not sure what was coming next. I suddenly started feeling the pressure; it was worse than the interview because now I was in front of six people I was going to be working with.)

Alex: Nurat, remember I had asked you in the interview if you have ever worked on Google Auto suggest box or not. You mentioned you hadn't, so now why don't you try doing a small demo for all of us on how you would automate such a control.

(Saying that he gave me the laptop which was connected to the projector and everyone could see exactly what I was going to do.)

One Month Later

Without wasting any time I just opened QTP and the Google website on IE. I did a view source on the web page, but could hardly find anything that may be used to implement the auto suggestion list. I said to myself, no matter what I need to get this done. I decided to view this in Firefox instead. And I launched FF and realized that there is no FireBug installed. I needed this to see the DOM tree.)

I need FireBug tool in FireFox to check the DOM implementation, can I download and install quickly on this laptop?

Alex: Sure, go ahead.

(All eyes were focussed on the screen noting down every move of mine.)

I downloaded the tool and installed it. Now I was trying to view the tree and see the changes that happen when we type something in the search text box.

By typing some text and clearing it again I found the node that was getting changed.

```
<div id="xjsd">
<div id="xjsi">
<script>
<div class="gac_od" style="visibility: hidden; left: 419px; top: 241px; min-width: 512px;">
    <div class="gac_id">
        <table class="gac_m" cellspacing="0" cellpadding="0">
            <tbody>
        </table>
    </div>
</div>
/span>
```

I got the lead into what I needed to do next. I typed in some text in the search box and noticed that the table with class="gac_m" was getting updated with search suggestion. I knew what I had to do next.

It took me another five minutes to write some code. Here is the code that will select and print the 3rd item from the list.

And I thought I knew QTP!

```
'Simulate typing
Setting.WebPackage("ReplayType") = 2

'Type some text into the search textbox
Browser("Google").Page("Google").WebEdit("q").Set "Tarun Lalwani"

'Wait for 500 msec
Wait 0,500

Dim oList
'Get the suggestion list table
Set    oList   =    Browser("Google").Page("Google").WebTable("html
tag:=TABLE", "class:=gac_m").Object

'Get the text of the 3rd item
Print "Item at 3rd position is - "& oList.rows(2).outerText

oList.rows(2).click
```

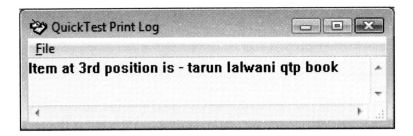

Alex: Can you explain to everyone what exactly is happening here?

Me: Sure. When we type anything in the search textbox there is hidden DIV element

```
<div class="gac_od" style="visibility: hidden; left: 419px; top:
241px; min-width: 512px;">
```

When we type some text into the search box this DIV element gets updated as

```
<div class="gac_od" style="visibility: visible; left: 419px; top:
241px; min-width: 512px;">
```

This DIV contains the table which gets updated with the suggested options

```
<table cellspacing="0" cellpadding="0"class="gac_m">
<tbody>
<tr class="gac_a"><td class="gac_c" style="width: 500px;">tarun lalwani<b> qtp book free download</b></td></tr>
<tr class="gac_a"><td class="gac_c">tarun lalwani<b> qtp book</b></td></tr>
<tr class="gac_a"><td class="gac_c">tarun lalwani<b> qtp ebook</b></td></tr>
<tr class="gac_a"><td class="gac_c">tarun lalwani<b> qtp</b></td></tr>
<tr class="gac_a"><td class="gac_c">tarun lalwani<b> qtp unplugged</b></td></tr>
```

This table has class as 'gac_m' and is unique in the source. So we can identify the table using the class and html tag property. Once we have access to the table we can get text of the row, as well, as perform a click on the same.

Alex smiled and everyone in the room started clapping. It really made me happy that I was able to overcome the first challenge that was put forward to me by my team. The exercise did help me become a part of the team from the first day itself and I started enjoying the work. The design for the framework had already started and there were lot of explorations being done. The team used to meet daily at 4:30 p.m. in the evening and discuss any problems or issues. Over the next few days I learnt lot of new things in QTP and framework designing. I was a silent spectator in most of these meeting, not because I didn't know about things but because I was very seriously listening to even the smallest of issues, making sure I didn't miss anything.

Today's meeting was to discuss about exploring the possibility to call a Function from global library and the Function will be present in the Action. Pankaj from the team was given the task and his opinion was that it may not be possible to do it because of QTP's architecture.

Alex: Nurat, what do you think?

(This would be the first time after my introduction meeting that I was about to speak.)

Me: I think Pankaj is right. Global Libraries are accessible to all Actions but Actions are not accessible to any of the global library code. So this is impossible in QTP.

Alex: Impossible is a very strong word Nurat! What do you think Andrew?

(Andrew was the person who had taken my 1st round of personal interview and had similar experience as mine.)

Andrew: I am not sure if it would be possible or not. But I will have to try few things before I can comment.

Alex: Fine, work on this and we will discuss this tomorrow.

The discussion moved on to other topics and today's meeting finally ended. I was not sure why Alex had said, "Impossible is a very strong word Nurat!" because I knew it was a very simple and obvious concept that it can't be done.

Next day the meeting started and I had already forgotten about the Actions question. We started the meeting with few new topics and finally came to the day before's Action question.

One Month Later

Alex: Andrew, did you find anything?

Andrew: Yes, there is a possible workaround for us to do this.

(I was shocked to hear this and started listening intently to what Andrew had to say.)

Andrew: Consider that we have a Function like PrintData which is specific to our Action

```
'Action1:
Call GlobalFunc
Function PrintData(a,b)
      'Print data in some format
End Function
```

```
'Global Library:
Function GlobalFunc()
      Call PrintData(4,5)
End Function
```

Now if we run the above code QTP would throw an error "Type mismatch" at Call PrintData(4,5).

(I was like... ofcourse that's why I said it's impossible.)

The reason for the same is that the global library cannot see the PrintData method. Now if we can give it a proxy pointer then the same would work. So we update the code like

```
'Action1:
'Give the print data proxy
```

```
Set fnGPrintData = GetRef("PrintData")
Call GlobalFunc

Function PrintData(a,b)
     'Print data in some format
End Function

'Global Library:
Dim fnGPrintData

Function GlobalFunc()
     Call fnGPrintData(4,5)
End Function
```

And now the global library Function will be able to call the Function in the Action.

(I started feeling a bit embarrassed that without giving it a proper thought I had declared that it was impossible. Suddenly it became clear what Alex had meant earlier. We all exited from the meeting room.)

Me: Nice solution, Andrew! None of that came to mind.

Andrew: Thanks!

It was around 6:30 p.m. I was at the rooftop cafeteria and could see the sun setting. I recounted the events of the day and smiling on my foolishness I said to myself, "And I thought I knew QTP!"

One Month Later

Lalwani

Acknowledgements

I would specially like to thank my family for this book. I took a 3-month sabbatical from my work to complete this book and my family extended every support to create a conductive environment and keep distractions at bay.

I would also like to thank Anshoo Arora (Founder of RelevantCodes.com) who has kindly extended his support as a Technical Editor and Reviewer of this book; Chhanda Burmaan who has contributed as an editor and ironed out the inconsistencies and fine-tuned the narration and Jophy Joy who made the book more interesting and lively with his humorous illustrations.

Anshoo Arora

Anshoo Arora is the founder of RelevantCodes.com — a blog targeted towards HP QuickTest professional. He is continuously researching to find better ways of working with QTP, simplifying automation maintenance, creating flexible test suites and frameworks. Apart from QTP, Anshoo also specializes in Quality Center, LoadRunner, .NET (C#, VB.NET) and Web technologies (HTML, CSS, JavaScript, and PHP). Anshoo can be contacted at: **RelevantCodes.com/contact**.

Chhanda Burmaan

After 10 years of experience in different functional areas across various domains like Publishing Outsourcing, Ebook Production, Banking, Telecom and IT, Chhanda is currently on a sabbatical to enjoy a more relaxed pace of life. In between vacations, she undertakes assignments that interest her and range from editing to sub-titling, scriptwriting and translation. She can be reached at **chhanda.burmaan@gmail.com.**

Jophy Joy

Jophy works as a UI (User Interface) designer and consultant. He has a great sense of humour matched with good capabilities to present them in visual form. Jophy has worked on the illustrations in this book and tried to capture the essence of the story at various junctures. He can be reached at **jtoonz@gmail.com.**

About the Author

Tarun Lalwani is a Test Automation & Solutions Architect and the author of the first ever book on QTP named "QuickTest Professional Unplugged". This is his 2nd book. He has worked on various automated testing projects utilising technologies like VBScript, VB6, VB.Net, C#.NET, Excel and Outlook Macros. He founded KnowledgeInbox.com — a blog targeted towards the QuickTest community. He uses the blog to share his custom APIs, products, articles, tips and tricks with his readers pro bono. He is also a regular contributor at AdvancedQTP and SQA forums. Tarun's work has been showcased on several websites such as:

 RelevantCodes.com

AdvancedQTP.com

Tarun was awarded with the 'Best Feedback Award' by HP for QTP 11 Beta testing. His book "QuickTest Professional Unplugged" was recognized as the Best Automation Book in the 2nd ATI Automation Honors award.

Tarun can be reached through any of the below links:

 KnowledgeInbox.com/contact-us

LinkedIn.com/in/tarunlalwani

QuickTest Professional Unplugged

QuickTest Professional Unplugged (now in its 2nd edition) is the first book written by author Tarun Lalwani and the first ever book on QTP as well. It has already turned out to be a best seller since its publication in 2009. Tarun Lalwani has won the Best Automation Book award in the 2nd ATI Automation Honors for the same.

This book is good for those starting out on a career in Testing Automation or even for those with a few years of QTP experience. It is the culmination of 3 years of research and effort in this field.

The book gives a pragmatic view of using QTP in various situations. And is recommended for those aspiring to be experts or advanced users of QTP.

Quotes from the Reviewers

"I find this to be a very pragmatic, hands-on book for those who want to extend their QTP skills beyond basic expert view programming. This book is written by a QTP master for those who wish to eventually become masters themselves." – Terry

"Tarun Lalwani has single-handedly helped thousands of people to expand their knowledge of QuickTest Professional. Here is a book the automated testing community has been crying out for. This book will help QTP practitioners, from beginners to experts. I have used QTP from V6.0 and during the review I learnt something from every chapter." – Mark

Where can you order this book?

In India:

Order online at **KnowledgeInbox.com/store**

US & Rest of the World:

QuickTest Professional Unplugged
www.lulu.com/commerce/index.php?fBuyContent=11253559

And I thought I knew QTP!
www.lulu.com/commerce/index.php?fBuyContent=10689195

For bulk orders & discounts, please email us at **orders@KnowledgeInbox.com**.

Your Feedback Counts!

We @ KnowledgeInbox take every feedback very seriously. Let us know what you think about this book — what you liked or what you didn't or simply what you would like to read about in a future book. Your feedback will help us in coming out with books that are relevant and in tune with what our readers want.

To send us general feedback, simply send an e-mail to **feedback@KnowledgeInbox.com** and please mention the book title in the subject line of your message. Alternatively you can leave a feedback online at **KnowledgeInbox.com/contact-us.**